D0975293

THE CRISIS IN ENERGY POLICY

THE GODKIN LECTURES AT HARVARD UNIVERSITY

The Godkin Lectures on the Essentials of Free Government and the Duties of the Citizen were established at Harvard University in 1903 in memory of Edwin Lawrence Godkin (1831–1902). They are given annually under the auspices of the John F. Kennedy School of Government.

THE CRISIS IN ENERGY POLICY

John M. Deutch

HARVARD UNIVERSITY PRESS

CAMBRIDGE, MASSACHUSETTS

LONDON, ENGLAND

2011

To my sons: Philip, Paul, and Zachary

Library of Congress Cataloging-in-Publication Data
Deutch, John M., 1938–
The crisis in energy policy / John M. Deutch.
p. cm.
"The Godkin Lectures on the Essentials of Free Government and the Duties
of the Citizen."
Includes bibliographical references and index.
ISBN 978-0-674-05826-2 (alk. paper)
1. Energy policy—United States.   I. Title.
HD9502.U52D48 2011
333.790973—dc22   2011010435

CONTENTS

ACKNOWLEDGMENTS

For thirty-five years I have worked on energy matters—as a government official, a university scholar, and an advisor to industry. During this time, I have been fortunate to form close personal and professional relationships with many talented and dedicated individuals. It is impractical to mention all who have informed and influenced my life, so I mention only a few.

I am grateful to James Schlesinger and Charles Duncan, the first two Secretaries of Energy, for their guidance and support during my service in the Carter administration as Director of Energy Research and Undersecretary. Both were unusual leaders who influenced my approach to public service. Jim Schlesinger, because of our common background in the Department of Defense and the intelligence community, has deepened my understanding of the international linkage between energy and security. As a member of the President's Nuclear Safety Oversight Committee, the President's Council on Science and Technology during the second administration of Presidents Reagan and Clinton, the Environmental Protection Agency Science Advisory Board, the Defense Science Board, as well as innumerable advisory committees and governing boards of Department of Energy laboratories, I have profited from the knowledge of leading experts in every conceivable energy technology.

At MIT, for more than a dozen years, I taught "Application of Technology to Energy and the Environment" with my colleague and friend Richard Lester. The purpose of this course was to introduce students to the complex task of integrating technical, economic, and environmental considerations required to understand energy issues. The enthusiasm of the students—who were often for the first time learning about the problems of applying technology outside their disciplines—as they grasp the challenge, has been enormously gratifying.

During more than forty years at MIT, I have worked with colleagues across the Institute on a wide range of energy problems. My collaborations have been stimulating, productive, and fun. Paul Joskow, for example, has wisely and consistently kept my technical aspirations in line with sensible economic reality; his policy judgment is unsurpassed. Ernie Moniz has been my closest technical collaborator for fifteen years; he has my admiration for his successful leadership of the MIT Energy Initiative. The MIT faculty has many talented and engaging scholars, who along with MIT students are the reasons why being a member of this university community is so rewarding.

I have also enjoyed a long and rewarding association with Harvard—Graham Allison, Joe Nye, L. Mahadevan, and George Whitesides each have had, and I hope will continue to have, a strong influence on my personal and professional life. Harvard's Kennedy School of Government has my thanks for the privilege of being the 2010 Godkin lecturer.

My interactions with industry have made me a better teacher, researcher, and public official. I deplore the stereotypes of ivory-towered academics who work on ideas and practical business men and women who solve problems. Leaders in both professions recognize that successful resolution of major societal technology problems requires the best new ideas, implemented in the most efficient manner possible. As a board member of Schlumberger, CMS Energy, ANR Pipeline, Cheniere Energy, CITIGROUP, and more recently as an advisor in several energy start-up companies—SION, Foro, MC-10, and Sun Catalytix—I have had the satisfaction of applying my technical and academic knowledge, and observing innovation happen in different practical situations. During my thirty-plus year association with Warburg-Pincus, I have learned a great deal about private investment in energy enterprises around the world, while deepening my personal and professional relationship with Harold Brown, former Secretary of Defense, Director of Lawrence Livermore National Laboratory, and President of Caltech, whose record of public service has been a model.

In sum, I am grateful for the combinations of factors—mentors, colleagues, and opportunities—that have permitted me to engage with our nation's energy challenges.

INTRODUCTION

Without energy, there is no economy. Without climate, there is no environment. Without economy and environment, there is no material well-being, no civil society, no personal or national security.

—John P. Holdren, "Energy for Change," *Innovations* 4, no. 4 (2009): 3

There are many pressing challenges facing the United States and other nations—economic growth, health, poverty, education, and energy. The energy challenge is how best to satisfy reasonable demand with secure, affordable sources of supply, without unduly harming the environment.

Despite recurrent energy crises and multiple calls for action, the United States fails to adopt consistent domestic energy policies and lags in taking a leadership role in international energy issues. Why this failure? What are the limitations in the way the U.S. government works that inhibit progress? What could the government do to work better?

My purpose is to answer these questions for readers who are familiar with public policy but not experts on energy matters. The invitation to deliver the 2010 Godkin Lecture at the Harvard Kennedy School seemed to be an ideal opportunity to assess how well our nation is preparing for our energy future. My views have been formed from my experience as an official in the U.S. Department of Energy in the Carter administration, and as director and advisor to several large energy concerns and technology start-ups. This book draws heavily on the thinking, teaching, and writing on energy issues that I have been doing at MIT over the past thirty-five years.

I hope to make three key points:

- The United States has failed to adopt and sustain an energy policy over the past decades.
- Energy policy is especially demanding because of (1) the linkage of domestic and international considerations in almost all energy issues; and (2) the requirement to integrate successfully technical, economic, and political factors in order to implement any significant change.
- There must be substantial changes in executive and legislative branch approaches to setting energy policy and administering energy programs, in order to prepare responsibly for the country's energy future.

I hasten to say that my intention is not to raise expectations that these changes are easy to achieve or imminent. The unlikelihood of success reflects deficiencies in the U.S. political process that inhibit addressing complex public policy issues that extend far beyond energy. However, it is preferable to have clearly in mind what needs to be done and to maintain some decorous level of anger that there is no change, rather than to accept the shortcomings of public administration with a worldly, pragmatic, but fatalistic view that our political system inevitably leads to less-than-perfect outcomes because of the need to balance and compromise different interests to make decisions.

Fundamentally this book is about the failure of rational policymaking to improve the nation's energy future. Instead over the past four decades political policymaking has dominated the country's approach to energy issues. There is an inevitable tension between rational policymaking and political policymaking. The former seeks to inform the decision maker by analysis of the costs and benefits of alternative courses of action to improve the public good. The latter proceeds through a process of negotiation and compromise to a result that creates winners and losers, more or less according to empowerment. Rational policymaking is not inevitably wise or efficient, and political policymaking is not inevitably venal or inequitable. However, for public issues such as energy that require consistent policies on a time scale that is long compared to election cycles and public attention, and changes that involve significant adjustment to current commercial practice and use of technology, the political approach is doomed—because the actors and their interests change too rapidly and there is little common understanding of the technical, economic, and international framework for energy issues.

Outline of the Book

Chapter 1 reviews the record of U.S. energy policy over the past three decades and identifies four causes of failure.

Chapters 2–4 examine analytic and policy choices in selected energy areas. These chapters are not intended to be comprehensive. Rather, I have selected issues that have commanded my attention over time, and therefore provide some basis for the opinions I express. In each instance, I seek to identify the gap between conventional wisdom and the underlying technical reality. Readers will find an unusual level of technical discussion for a book devoted to policy. I do not apologize for this intrusion, given that political and policy deliberations on energy issues must internalize, at some level, the technical dimension.

Chapter 2 is devoted to climate change, one of the most complex and vexing energy issues. The climate issue has dominated the U.S. energy policy debate since the United States failed to ratify the 1997 Kyoto Protocol, despite signing the UN Framework Convention on Climate Change in Rio de Janeiro in 1992 during the George H. W. Bush administration. The climate

issue illustrates the complex interaction between selecting technical options to reduce emissions, crafting a policy that balances
differing political interests with economic incentives that will
encourage change, and negotiating an international agreement
that harmonizes the policies of developed and developing countries. I discuss the approach that President Obama has taken toward climate change in his first term and its shortcomings.

Chapter 3 addresses the key energy security issue of oil import dependence. The public's attention was drawn to energy
originally by a series of energy security events in the 1970s discussed in Chapter 1: possible Soviet intervention in oil-producing
countries of the Middle East, India's explosion of a nuclear device, and especially lines at gasoline stations created by oil supply disruption in 1973 and 1979. So there has long been public
interest in the United States' becoming "energy independent"
from oil imports. Advocates on the left and right have shame-
lessly exploited this understandable and reasonable public aspiration, in arguing for government subsidies for their preferred
solution.

In Chapter 3 I seek to establish two points. The first is that
the security cost of oil import dependence is not the economic
cost that the economy might bear from a supply interruption.
If it were, compensating policy measures would be relatively
simple—for example, building "strategic petroleum reserves"
designed to dampen the effects of interruption, and extending
a "security premium" to domestic and non-OPEC oil producers.

However, the security cost of import dependence is of a very
different nature. Import dependence constrains the leverage of
the United States in all foreign policy matters, because our
allies and adversaries understand that every foreign policy measure must take into account the possible effect on oil imports in
particular and global world oil trade in general. Moreover, major oil resource holders, as well as developed and developing

countries that rely on oil imports, are tempted to reach bilateral arrangements to avoid the uncertainties of a transparent global oil market. Such bilateral relationships tend to replace economic competition with political alliances. Because many of the major resource holders are countries whose interests and values are not aligned with those of the United States, oil dependence sharply affects U.S. foreign policy interests. Accordingly, it is natural to seek policies that will make the U.S. independent of oil imports.

The second point I make in Chapter 3 is that effective energy independence is not a realistic goal for U.S. energy policy. Instead, the United States must prepare to manage the consequences of our and our allies' oil dependence in our foreign policy.

Chapter 4 examines the prospects for three key technologies—solar, biomass, and nuclear—that are expected to be increasingly important in the future because they are largely carbon-free. I stress both the opportunity and the challenge of deploying new energy technologies. The gap between aspiration and technical and economic reality is large in each case, but for different reasons.

Chapter 5 deals with the government's role in encouraging and managing new energy technologies. I often hear that technology is the answer to the energy problem. The argument is that if the United States could send a man to the moon and make the atomic bomb, an organized energy technology effort similar to the space program or the Manhattan Project should produce breakthrough technologies that solve the energy problem. I believe the search for magic-bullet technical solutions is futile. There will be multiple innovation paths that influence both energy supply and demand.

Is technology the answer? The answer is almost certainly no, in the sense that unforeseen technical breakthroughs will not create ways to produce inexhaustible, inexpensive sources of energy that have no adverse environmental effects. The answer is almost certainly yes, in the sense that new energy technologies will, over the long term and at significant cost, replace existing technologies.

The pace of this transformation depends both on the foresight of government policy and on the management of government and industry technology development efforts. Chapter 5 discusses how the federal government has been managing technology, how it should be managed, and importantly the government's relationship to private sector innovation activity.

In Chapter 6, the final chapter, I propose changes that I believe are needed in how the United States approaches energy issues. Change is needed in the executive branch, most especially the Department of Energy, in the Congress, and in the balance between state and federal responsibilities. Most observant readers will note that the proposed changes are unrealistic, because the changes fundamentally require a rebalancing between local and private interests and the public good. In the American system, such change occurs, if at all, very slowly or in response to genuine crisis. My point, however, is not that the proposed changes are likely to occur, but rather that the historical evidence and the nature of the energy problem are such that if change does not occur, there will continue to be no progress. The result will be that future generations will bear greater economic costs, social dislocation, and possible conflicts than is necessary.

1

THE FAILURE OF U.S. ENERGY POLICY

In this chapter I briefly review the emergence of energy as an issue for the United States in the 1970s. I identify four fundamental reasons why the U.S. government has been unable to formulate and sustain an effective energy policy: (1) the adoption of popular but unrealistic goals, (2) public attitudes and underlying moral dilemmas, (3) the complication of competing domestic and international considerations, and (4) the absence of quantitative analysis in planning, policymaking, and administration of government programs. I conclude by contrasting the approaches to energy policymaking of Presidents Carter and Obama, two leaders genuinely committed to ensuring the energy future of the country.

Two Energy Crises of the 1970s Raise the Alarm

In October 1973 the twelve members of the Organization of Petroleum Exporting Countries (OPEC), joined by Egypt, Syria, and Tunisia, announced an oil embargo in response to United States assistance to Israel during the Yom Kippur War. The subsequent cut in production of about 25% created a supply interruption and caused the price of oil in the United States to jump from $3 per barrel to $12 per barrel in 1974 dollars (over $60 per barrel in 2008 dollars). The immediate effect was an oil shortage and lines at gas stations; the embargo continued until March 1974. President Nixon sought and received authority from Congress for price, production, and market allocation controls; in November 1973 he launched Project Independence (discussed below).

The embargo defined "oil as a weapon" for the United States and other major importing countries. Streams of revenue flowed from the major consuming countries to the major producing countries, especially to the politically unstable and not entirely friendly Middle East. The embargo showed that OPEC, by restricting production, had the power to disrupt markets and impose economic hardship on the United States and other importing countries. In fact, in the second oil crisis of 1979, caused by the loss of Iranian oil production during the Iranian revolution, OPEC expanded production in an attempt to dampen the panic and soaring oil prices, but lines again formed at gasoline stations and were not alleviated by clumsy U.S. government efforts to allocate available supplies to meet shortages around the country. Totally impractical schemes for rationing motor gasoline were seriously discussed, and some rationing coupons were even printed (see Figure 1.1).

The tremendous increase in the real price of oil led to an explosion in the value of Middle East oil reserves. In the height of the Cold War, the presence of massive conventional Soviet military forces across the Caucasus and the 1979 Soviet invasion of Afghanistan heightened concerns about possible Soviet

Figure 1.1.
A rationing coupon.

aggression against Persian Gulf nations. President Carter stated that the United States would use military force to defend its interests in the region (see Chapter 3).

These two oil crises defined for the U.S. public the economic and security disadvantages of oil import dependence and established the first important principle guiding U.S. energy policy: reduce dependence on imported oil and begin a long transition from an economy based on increasingly expensive oil.

Not surprisingly, the sharp increase in the real oil price led, beginning in 1980, to a steep reduction in demand, especially for motor gasoline, and the price of oil declined from a high of over $90 per barrel to a low of $20 per barrel in 1986 (in 2008 dollars). The second important principle of energy policy was established in the public's mind: industry and individual consumers will reduce energy consumption as energy prices increase; accordingly prices should be set by the market and not subsidized.

Nuclear Power and Climate Emerge as Issues

Of course, not all energy comes from oil and gas. Nuclear power and coal are important sources of supply. In 1974 India's explosion of the 8-kiloton "Smiling Buddha" plutonium implosion bomb at Pokhran in the Rajasthan desert, described by Indian authorities as a "peaceful nuclear device," raised concerns about proliferation and the possible misuse of nuclear power to bring nations closer to a nuclear weapons capability.

The previous six countries that possessed a nuclear weapons capability had acquired the highly enriched uranium or plutonium required for the bomb from covert facilities, such as reactors dedicated to plutonium production for weapons. India's plutonium came from a 40-megawatt thermal CIRUS heavy-water reactor acquired from Canada. Nuclear weapons material is either uranium that is highly enriched in the fissile U^{235} isotope, which is only present at 0.7% in natural uranium, or plutonium, which is produced by neutron absorption from irradiation of the abundant U^{238} isotope in a reactor and chemically separated from the irradiated nuclear fuel after discharge. Enrichment and reprocessing are fuel cycle processes that accompany the production of nuclear power.

In the early 1970s several nuclear supplier countries, including the United States, the United Kingdom, France, Germany, and Japan, were considering the export of enrichment and reprocessing plants to other countries, including Brazil, Argentina, South Korea, and Taiwan. The Indian explosion drew attention to the possibility that the spread of these sensitive technologies would bring many countries closer to a weapons capability. The United States took the lead in advocating international export controls to slow the spread of nuclear weapons, and reducing the risk of proliferation from commercial nuclear power became the third important principle of U.S. energy policy.

Public attention also turned to sustainability and environmental conditions, such as climate change. In 1963 the United States passed the Clean Air Act (CAA) to control air pollution by reducing the atmospheric concentration of pollutants to levels that the Environmental Protection Agency judged safe for human health.[1] Amendments to the CAA in 1970 designated six "criteria pollutants": ground-level ozone, particulate matter (PM_{10}), carbon monoxide (CO), sulfur oxides (SO_x), nitrogen oxides (NO_x), and lead.[2] The 1990 amendment to the CAA authorized firms to purchase and trade sulfur and nitric oxides emission permits, creating a market mechanism to achieve the least-cost compliance with the established cap. This "cap and trade" program is uniformly judged to have been a great success in lowering the overall cost of compliance in the United States and Europe.

This success has led many experts to favor a cap-and-trade system as the preferred mechanism for reducing global emissions of greenhouse gases (GHGs), as compared, for example, to an emissions tax. However, there are several important differences between these two applications. The sources of the CAA criteria pollutants are largely in the jurisdiction that benefits from reductions. This is not true for GHGs, as global emissions determine local effects. Second, the sources of SO_x and NO_x are mainly major fuel-burning installations, such as coal-burning power plants, so that a system can be successful by placing emission caps on a relatively small number of emitters, greatly facilitating enforcement. Third, the residence time of GHGs in the atmosphere is on the order of one hundred years, while the residence time for air pollutants such as SO_x is on the order of one year. This long time scale for GHG lifetime in the atmosphere creates an equity issue between the developed countries, whose past emissions are responsible for a greater proportion of the current inventory of GHGs in the atmosphere, and the developing countries, whose emissions will grow in the future. In sum, a cap-and-trade system for global GHG emission reduction is a good deal more difficult to craft than one for air pollution, a much more local phenomenon.

Concern about the influence of carbon dioxide in the atmosphere on atmospheric temperature dates back to the Swedish Nobel laureate Svante Arhenius, in a paper he presented to the Swedish Physical Society in 1895. In 1965 the Environmental Pollution Panel, in the influential report *Restoring the Quality of Our Environment,* prepared for President Lyndon Johnson by the President's Science Advisory Committee, signaled the importance of carbon dioxide in the atmosphere.[3] In the early 1980s the scientific community expressed heightened concern that the increase in GHGs from anthropogenic sources threatened to tip the delicate natural radiative balance toward higher global temperatures due to the "greenhouse effect." The GHGs in the atmosphere absorb outgoing long-wavelength radiation, preventing the radiation from escaping, and thus heating the planet.[4]

The United Nations established the Framework on Climate Change (UNFCC) mechanism at the Earth Summit in Rio in 1992. The work of the UNFCC is supported by a large and impressive International Panel on Climate Change (IPCC) that brings together hundreds of qualified scientists from many countries to address various aspects of the climate problem. The IPCC has issued four general assessment reports (in 1990, 1997, 2001, and 2007), as well as many special reports.[5]

The business of the UNFCC is transacted at Conferences of the Parties (COPs). At the third COP in Kyoto in 1997, parties agreed to the Kyoto Protocol: 37 industrialized countries agreed to binding reductions in GHGs emission of 5% below 1990 levels by 2008–2012. The United States signed the Kyoto Protocol but did not ratify it. The protocol adopted no reduction targets for developing countries. The fifteenth COP, held in Copenhagen in 2009, was intended to be the occasion when agreement would be reached on binding reduction targets that would be accepted by the developing economies. Agreement was not reached, but disaster was averted by the dramatic intervention of President Obama in a meeting of China, India, Brazil, and South Africa that led to some concessions.[6] The sixteenth COP was held at Cancun in late 2010; expectations were lower, and some significant progress was made. As of this writing (January 2011), there is no clear path forward to indicate how the developed and developing countries will agree to limit emissions at the seventeenth COP to be held in Durban, South Africa, in December 2011.

The point is that the major items have been on the energy agenda of the United States for a long time. For over two decades it has been clear that the United States should

- adopt an effective climate change policy and lead the world in reducing the risks of climate change,
- begin the long transition from an economy based on fossil fuels to one reliant on renewable energy technologies and nuclear power, and
- continue to make progress on increasing the efficiency of energy use and reducing oil import dependence.

There are at least four reasons why it has proven so difficult for the United States to adopt a stable and effective energy policy:

1. Setting unrealistic goals. Since the first oil embargo of 1973, it is not surprising that U.S. presidents, political leaders, and candidates have set energy goals. In part this is a genuine expression of leadership; in part it is an unavoidable tendency of national leaders to say what the public wants to hear.

When in November 1973 President Nixon, in response to the OPEC oil embargo, announced Project Independence, he declared: "Let this be our national goal: At the end of this decade, in the year 1980, the United States will not be dependent on any other country for the energy we need to provide our jobs, to heat our homes, and to keep our transportation moving."[7] The goal was not met. The Energy Information Administration (EIA) of the U.S. Department of Energy (DOE) estimates that in 2010, imported petroleum will be about 50% of the U.S. liquid fuel consumption, compared to about 20% in 1973.[8]

In June 1979 President Jimmy Carter announced a solar strategy and set a renewable energy goal of 20% of U.S. energy consumption for the year 2000:

> This solar strategy will not be easy to accomplish. It will be a tremendous, exciting challenge to the American people, a challenge as important as exploring our first frontiers or building

the greatest industrial society on Earth. By the end of this century, I want our Nation to derive 20 percent of all the energy we use from the Sun—direct solar energy in radiation and also renewable forms of energy derived more indirectly from the Sun. This is a bold proposal, and it's an ambitious goal. But it is attainable if we have the will to achieve it.[9]

This goal was not met. The total renewable energy contribution to U.S. energy supply remains remarkably close to the 9% level it was at when President Carter made his speech in 1979.

I am especially sensitive to this second example, because at that time I was a DOE official in the Carter administration, and I was called on to answer questions about the so-called White House "fact sheet" that explained how this goal might be reached.

The implication of setting goals deserves several comments. First, energy pronouncements that are clearly designed for a domestic audience often have adverse international consequences (and vice versa). Our allies, such as Germany and Japan, who will remain dependent on oil and natural gas imports even if we were able to reduce U.S. imports to zero, do not view U.S. energy independence as constructive, because the difference in import dependence could create strains in the alliance.

President George W. Bush in his 2006 State of the Union message proposed a goal of 75% reduction in oil imports from the Middle East by 2025.[10] This statement prompted an astonished high-level Saudi official to question the credibility of President Bush's contemporaneous request to the Saudis that they increase oil production to help dampen the sharp rise in oil prices. Domestic goals often sound peculiar to foreign ears.

President Obama has consistently expressed his support for recent climate legislation that mandates a 17% reduction in greenhouse gas emissions by 2020 and 80% by 2050. The assumption is that the mandate will cause market adjustments that ensure compliance. However, attempts to model[11] how energy markets might adjust to meet the legislated constraints need to make heroic assumptions[12] about reductions in energy intensity as well as the technology readiness, cost, and demonstrated availability of alternative carbon-free technologies (such as carbon capture and sequestration, and cellulosic ethanol for transportation fuels) to meet such goals. In Chapter 2, I examine in some detail what sensible models tell us about the uncertainty of achieving future emission reductions.

Why do political leaders set energy goals? Certainly leaders believe that aspirational goals are important for defining a "vision" of the future that will enlist public support and catalyze change. This is true in areas other than energy. Leaders surely intend for the aspirational goal they espouse to result in practical advances. In some cases where unrealistic goals have been set, significant progress has occurred—for example, in improving water quality—so it would be an overstatement to say that unrealistic goals or targets can never lead to progress.

But I observe in the energy area that leaders tend to avoid speaking realistically about the choices and sacrifices needed to make progress. They prefer to set politically popular goals that reflect the aspirations of the American people, who understandably want cheap energy that is not environmentally harmful and not dependent on imports from unstable and unreliable parts of the world. Political leaders find it difficult to avoid the temptation to set these goals without addressing the harsh realities of the resources, time, and higher costs required.

Unrealistic goals inevitably are reversed or ignored and make the public ever more cynical. False signals are sent to private investors and banks about when and where to extend credit and commit risk capital. Major corporations, such as Exxon, DOW, and BP, which invested in solar or synthetic fuels in the 1970s, are understandably reluctant to invest, absent more certain energy policies. Venture capitalists and start-up firms are left high and dry.

And, of course, the leaders who set the goals will likely no longer be in office to explain why the long-term goals were not met.

2. Public opinion and underlying moral attitudes. A great deal is known about the public's attitudes toward energy issues. Public opinion surveys over the period 1974 to 2006 show that public concern about energy has remained high since the energy crises of the 1970s.[13] Respondents term the energy situation "very serious," see the chance of energy shortages as increasingly less likely, and attribute the cause of energy "problems" (in order of importance) to oil companies, oil-exporting countries, the administration, the Congress, auto companies, and energy demand. Support for increased conservation measures and renewable energy is consistently high. Respondents have, over time, become more closely split on opening the Arctic National Wildlife Refuge to oil exploration.[14]

Opposition to nuclear power has declined in recent years, although every segment of the public[15] opposes construction of nuclear, coal, and natural gas electricity-generating plants in its own locality, except for wind farms.[16] Some surveys indicate that support is growing for public measures that deal directly with carbon emissions and climate change. For example, the public views climate change and oil dependence as equally compelling rationales for increased fuel taxes.[17] However, recent polls[18,19] over the past two years reveal that the public now believes that the seriousness of global warming is generally exaggerated.

My review of public opinion surveys on energy issues has led me to these conclusions: (1) A significant proportion of the public is deeply concerned about the country's energy future;

(2) energy costs, climate change, and energy security are the leading concerns in the public's mind today; and (3) the public is more open than previously to energy conservation measures and new technologies (except for, of course, a gasoline tax that would directly impact drivers' pocketbooks).

Among the fraction of the public that is particularly engaged with energy issues, there is a strong and committed grassroots movement dedicated to protecting the environment from energy projects and activities, such as drilling in the Arctic National Wildlife Refuge, nuclear waste disposal, and coal combustion. I detect a moral sentiment underlying the attitudes of this public interest community. Many believe that available energy is not fairly distributed to different elements of society. In the United States, the call for a "sustainable" energy economy, ostensibly expressed as seeking assurance that any energy action taken today will not cause future imbalance, often reveals a moral sentiment about the virtue of a simpler way of life, with less emphasis on economic growth and consumption.

In the United States the commitment to energy issues is particularly strong among university students. An impressive number of students at MIT, and in other schools I have visited around the country, elect to study energy-related subjects and aspire to a career in the energy field—whether in government service, a start-up, or a large energy concern. Why this trend? Students recognize the importance of energy to the world's political and economic future and happily possess an idealistic vision of what is possible to achieve. In choosing a career, students also have an uncanny ability to pick fields that offer opportunity—energy challenges are seen as long-term winners in both professional terms and service to society. Finally, the energy field opens a way for students to integrate their largely discipline-oriented studies—science, economics, politics, and international affairs—with a subject that will have practical importance for the society over many decades. Students' widespread interest in energy is one of the few positive trends on the energy landscape.

Public attitudes about energy are of course also important in other countries, and these attitudes can differ sharply from attitudes in the United States. For example, attitudes toward climate change differ dramatically between the developed world (the United States, the European Union, and developed Asia) and the developing world (China, India, and other rapidly growing economies in Asia, Africa, and Latin America). Attitudes differ about whether to adopt an efficiency principle (reducing greenhouse gas emissions per unit of GDP) or an equity principle (equal GHG emissions per capita) to guide national emissions limits. The extended debate at the Conference of the Parties to the U.N. Climate Change Convention held in Copenhagen in 2009 makes it clear that this is partially an issue of economic interest and partially a moral difference about equity among past, present, and future generations' right to emit greenhouse gases.

Strong opinion backed by moral sentiment is often a desirable and effective way to mobilize action on a public issue. However, in other circumstances strongly held views reflecting different interests supported by moral sentiment can cause conflict. History is replete with both examples. Strongly held views often preclude finding new solutions or crafting workable compromise. As I am a bare-knuckles technical person, I cannot presume to offer any further insight into this dilemma.

3. The linkage between international and domestic energy policy. Energy issues are distinctive because international linkages dominate. In contrast, despite increasing globalization, other important issues, such as health care, have an essentially domestic character, in the sense that the benefits and the costs generally occur within the state or community that undertakes the action. Thus political decisions are largely determined by domestic considerations.

For energy and many related environmental matters, this is not the case: domestic policy decisions have international consequences, and international events influence domestic energy conditions. Progress on climate change, managing a transition from progressively more expensive imported oil upon which the United States and many other countries depend, requires change on a global scale. This situation belies the dictum of a former speaker of the House, Thomas "Tip" O'Neill, that "all politics is local" and instead signals the difficulties that arise from the requirement to balance domestic and international considerations.

The central point is that managing energy issues requires integration of international considerations, in addition to complex interrelated technical, economic, and political factors, and thus makes effective and acceptable solutions unusually difficult to attain.

There are exceptions, however, that illuminate the way forward. For example, over the past thirty years U.S. policy has recognized that responsible development of nuclear power requires integration of technology (power production, safety, and fuel cycles), economics (cost relative to coal, natural gas, and renewable generation), domestic politics (public attitudes toward safety and waste management), but also international risks (proliferation risk of uncontrolled expansion).[20]

The consequences of the complexity added by international considerations are well illustrated by comparing the success of acid rain legislation in the 1980s and the inability to pass climate change legislation today. The acid rain legislation, which reduced U.S. emissions of sulfur and nitric oxides through an effective U.S. cap-and-trade system,[21] was essentially domestic in nature (Canadian emissions transported across the border is the exception.). Effective climate change legislation, because it must be part of a global agreement on emissions reductions, inherently balances domestic and international considerations. And, of course, a global cap-and-trade system is much more difficult to design, implement, and verify.

4. Plans and numbers. Understanding energy issues requires integration of technical, economic, political, and international considerations. Introducing carbon capture and sequestration for coal-fired electricity generation, a smart national electricity grid that encompasses demand-side management, and new efficiency standards for buildings or appliances are complex technical subjects that exist in an economic and political context. This context is at least equally important as technology in designing a practical and efficient regulatory policy. It is madness to imagine adopting a policy without careful quantitative analysis of the costs and benefits of alternative actions to accomplish a designated purpose, and the trade-offs and uncertainties among these choices. Yet the U.S. Congress and executive branch frequently propose energy policies without serious analysis. One example, already mentioned, is the popular adoption of unrealistic and unsupported goals. A second example is the embarking on ambitious programs, such as modernizing the electricity grid of the country, without adequate modeling and simulation to identify the trade-offs between generation, transmission, and energy storage. A third example is adopting multiple measures to encourage renewable energy deployment without examining which one is most cost-effective. I shall describe in Chapter 4 my unhappy experience with the congressional initiative in the late 1970s to encourage use of gasohol—ethanol made from corn mixed with motor gasoline.

Quantitative analysis has value beyond avoiding needlessly expensive energy policies. Multiyear plans are an essential means of managing public programs. If results and resources cannot be measured on an annual basis, it is impossible for presidents, the Congress, or the public to know what is working. Furthermore, there are significant uncertainties—for example, the future cost of oil and gas and the rate of progress on waste management—where the outcome, once known, will influence future choices. Modeling informs decision makers confronted with uncertainty about the cost and benefits of options they should preserve for future action. I am consistently amazed at the lack of systematic effort the U.S. Department of Energy applies in modeling technical, economic, and regulatory aspects of various sectors, notably coal, nuclear, and electricity.[22]

Quantitative analysis does not determine decisions. The possession of sound analysis equips a policymaker to make a case. The absence of analysis leaves the decision outcome more vulnerable to the pressure of special interests. Les Aspin, former secretary of defense and my friend from our days as MIT graduate students, advised: "You must begin with understanding the analysis supporting a sound policy and then decide how best to craft a political strategy to accomplish it." Harvard professor Robert Stavins observes that policy analysis should be used as a "light bulb," not a "brick," in political debate. So while it is necessary to acknowledge Bismarck's dictum that "politics is about sentiment, not calculation," crafting policies that are both acceptable *and* cost-effective requires serious attention to numbers.

In Chapter 6, I offer two recommendations intended to redress the deficient attention to plans and numbers.

The Contrasting Approaches of Presidents Carter and Obama to Energy Policy

The record shows that despite the importance of the energy issue, the widespread public concern about energy, and the many expressions of necessary goals, the United States executive branch had been unable to agree on a path to securing the nation's energy future or legislation that establishes a comprehensive and enduring energy policy. I take *policy* to mean a comprehensive and coordinated multiyear plan of action to achieve adopted objectives, for all government agencies, supported by programs that are allocated adequate resources. The national energy policy should be accompanied by a plan that, to the extent possible, is based on quantitative analysis of the costs and benefits of different elements of the plan. The analysis should identify measurable benchmarks of progress and provide for the evaluation of different courses of action in response to changing conditions.

Different presidents have brought significant philosophical differences and approaches to the energy problem. President Carter, for instance, characterized the challenges as the moral equivalent of war:

> Our decision about energy will test the character of the American people and the ability of the President and the Congress to govern. This difficult effort will be the "moral equivalent of war"—except that we will be uniting our efforts to build and not destroy.[23]

President Reagan, on the other hand, evidently preferred to rely on the market and the private sector to meet energy needs. Only Presidents Carter and Obama have been strong advocates for the need for national energy legislation, but their approaches have been very different, and it is instructive to compare the approach of President Carter in the 1970s with the approach of President Obama today.

In April 1977, President Carter laid out an admirable energy policy that described the challenges clearly.[24] He proposed ten policy principles, and then stated:

I am sure each of you will find something you don't like about the specifics of our proposal. It will demand that we make sacrifices and changes in our lives. To some degree, the sacrifices will be painful—but so is any meaningful sacrifice. It will lead to some higher costs, and to some greater inconveniences for everyone.

Carter proposed to establish the Department of Energy (DOE) and then set two important objectives for the new department in its organization act:[25]

(2) To achieve, through the Department, effective management of energy functions of the Federal Government . . .

(3) To provide for a mechanism through which a coordinated national energy policy can be formulated and implemented to deal with the short-, mid- and long-term energy problems of the Nation; and to develop plans and programs for dealing with domestic energy production and import shortages.

Simply put, the DOE was to manage the energy programs and formulate a *national* energy policy. Few would deny that the DOE has not come close to achieving these objectives. Except for the National Energy Plans of 1978 and 1979, I suggest, the DOE has not offered a comprehensive national energy plan based on analysis.[26] The Energy Information Administration (EIA) of the DOE has done an admirable job of projecting domestic and international energy statistics.[27] But statistics alone do not constitute an energy plan that lays out the consequences of different policy choices, based on modeling and simulation, for various energy development paths.

Carter presented the Congress with detailed proposals and worked with Congress to enact comprehensive energy legislation—the National Energy Act of 1978 and the Energy Security Act of 1979.[28] During the Reagan years, Congress revoked most of the provisions of this legislation.

President Obama has taken a different approach. Rather than present Congress with a framework for climate change, he has asked Congress to craft legislation. I believe a legislative strategy that delegates to Congress an issue that has complex technical, economic, and political aspects is equivalent to shouting "jump ball" and will inevitably result in faulty and inadequate legislation. This has certainly proved to be the case with climate legislation in the Obama administration.

The Waxman-Markey House bill and the Boxer-Kerry Senate bill, 1400- and 800-plus pages respectively, were Christmas-tree-like bills that included a bewildering and contradictory array of measures assembled to attract sufficient votes for passage. If Congress had enacted either of these two approaches or some alternative, such as the Kerry-Lieberman bill[29] (which also proposed 80% GHG reduction by 2050), the result would have been too weak to launch the country on a much-needed new energy path, as I will explain in a subsequent chapter.

Climate change advocates sheepishly maintain that imperfect legislation on tough issues is the inevitable consequence of the process of compromise endemic to our system, and that the legislation would be a "first step" toward dealing seriously with the climate problem; perhaps so. But again the record shows that the U.S. Congress, reflecting regional differences and intense pressures from industry, local government, and public interest groups, is unable to adopt national legislation and keep it in place for sufficient time to have the necessary effect on reshaping the U.S. energy economy. The legislative strategy of taking a "first step" might be reasonable, if there were some understanding of the content and timing of a "second step" to address remaining climate issues—unfortunately there is no such understanding.

The President as Chief Executive or Chief Negotiator

Presidents Carter and Obama share similar objectives but very different approaches to convincing Congress to pass legislation to support their programs. President Carter followed an executive approach that provided draft legislation to Congress, quite broad in character, supported by a national energy plan based on numbers. Carter put forward not only ideas but also specific implementation proposals for the National Energy Plan. Congress passed much of Carter's 1978 legislative initiative, although with many changes required to gain legislative support. The strength of this executive approach is that the president defines the starting point for legislative dialogue and a framework that constrains to some extent the natural legislative tendency to craft a bill that reflects the interests of senior members. This auspicious start did not lead to a sustained energy policy for the United States. Carter was a one-term president. His successor, Ronald Reagan, who had a free-market approach to energy, was easily able to dismember the Carter energy program as oil prices plunged in the early 1980s.

President Obama follows a negotiation approach to energy legislation. In his first term he made it clear that climate legislation was a priority, and he made clear the elements he wanted to see in legislation. But Obama did not offer an administration climate proposal or a plan supported by numbers that credibly would achieve the GHG emissions he believed necessary. Rather, Obama responded to legislation passed by the House and Senate by negotiating with legislators to change specific aspects. This approach reflects confidence in the ability of Congress to draft legislation without an explicit programmatic structure to achieve desired energy objectives. It is an approach vaguely reminiscent of Ronald Reagan—the president as leader, pointing the way forward, rather than as an executive, responsible for implementation of a coherent program that may or may not succeed.

Perhaps President Obama in his second term will convince the country to adopt a long-overdue effective national climate policy, but this is far from certain, and the consequences of inaction are serious.

The Result of Inaction at the National Level on Energy Policy

There are three consequences of inaction at the national level. First, important industries, such as public utilities, are left uncertain about future government regulations covering a wide range of subjects: carbon emission charges, renewable electricity portfolio standards, domestic drilling for oil and gas, nuclear waste disposal, energy efficiency standards for buildings, automobile fuel economy standards, and so on. Private concerns, of course, routinely must accept a good deal of uncertainty in their investment decisions about market demand and price. But persistent and significant uncertainties about domestic policies in the long run serve only to slow investment and to encourage firms to move to other jurisdictions with more favorable, or at least more certain, policies.

Second, the public demands action. If action does not take place at the national level, the public will expect action at the local level. States and regions will adopt differing policies that meet local interests. For example, California's Assembly Bill 32, modestly entitled The California Global Warming Solutions Act of 2006 (AB 32), is the cornerstone of California's initiative to reduce GHG emissions to 1990 levels by 2020 and to further reduce GHG emissions to 80% of 1990 levels by 2050.[30] Key regulations must be implemented by the California Air Resource

Board to support attainment of these goals, which will include requirements for a cap-and-trade system for GHG emissions from utilities and other large combustion facilities and a low carbon fuel standard for transportation fuels. Indiana subsidizes advanced coal conversion projects that will burn coal, locally mined, more efficiently and with lower emissions. The northeastern states have adopted a Regional Greenhouse Gas Initiative (RGGI) to establish a regional cap-and-trade system to control CO_2 emissions by allocating emission allowances that may be traded between generators.[31]

The result is a patchwork of inefficient energy regulations that will prove difficult to reconcile with national legislation, whatever might be enacted. It is noteworthy that the European Union is moving in exactly the opposite direction—adopting more uniform regional energy rules—and that most energy experts would be aghast if China or India were to pursue such a local approach in managing their energy futures.

Third, absent congressional action, the courts and an increasingly impatient administration will take action with whatever administrative tools are available. On April 2, 2007, the U.S. Supreme Court ruled that the Environmental Protection Agency was required to regulate GHGs under the Clean Air Act. The Obama administration has made it clear that it will use the CAA authority to reduce GHG emissions. The EPA has made an "endangerment finding" that GHGs are harmful pollutants under the CAA that threaten the public health and welfare.[32] The agency is moving to issue mandatory regulations governing emissions that are unlikely to prove as efficient as a market-based mechanism. I doubt that use of regulatory mandates will prove to be an effective way to establish national energy policy for climate change, much less other important energy issues. The EPA has neither the authority nor the expertise to develop an approach that balances economic and climate considerations along with international implications and technology development. Furthermore, subsequent administrations will find it easier to dismantle regulations than to reverse well-thought-out national energy legislation.

Over the past forty years there has been some progress on energy issues: energy prices have been deregulated, air quality has been improved through the use of an efficient market mechanism, energy productivity has steadily increased, and the international linkage of energy issues is more widely appreciated. But little progress has been made on the major issues, and this points to the urgent need to make structural change in how the United States is approaching energy policy.

2

ENERGY AND CLIMATE CHANGE

In the past decade climate change has become perhaps the most important and contentious energy issue. Continuous anthropogenic emission of greenhouse gases (GHGs) increases the concentration of GHGs in the atmosphere.[1] The increase in concentration, in turn, increases global temperatures, which affects climate in a manner that leads to adverse effects on the environment and economic and social activity.

There are three different approaches for avoiding these adverse effects: (1) reducing global emissions, (2) learning to adapt to the changes in climate, or (3) taking active measures to counteract the warming trend. For the past two decades, international effort has been devoted to the first approach—reducing emissions. The key issues that need to be addressed are these: the validity of the science, the measures that should be taken today, and, most importantly, how the cost of these measures should be shared between developed and developing countries. In brief, the current status is that although there is a consensus on the scientific basis of climate change, questions continue to be raised. Despite extended debate, the United States has failed to adopt a comprehensive climate policy, which all agree is a necessary but not sufficient condition for reaching international agreement on reducing global emissions. The 1992 Kyoto Protocol placed emphasis on reducing GHG emissions, but without agreement on the extent to which these reductions should be shared between developed economies, such as the United States, which have been responsible for the bulk of past emissions, and the developing economies, such as China and India, which are expected to be responsible for the bulk of future emissions.

Climate is a perfect example of the type of public policy problem that challenges governments: a complicated technical problem whose solution requires vast resources, international cooperation, and consistent actions by industry and government over many years.

So far, acceptable progress has not been made on climate policy. The United States has failed to pass climate legislation, which in any case, in all likelihood, would have been too weak to inspire confidence that U.S. emissions will reach the ambitious midcentury goals established by the legislation and as announced by President Obama.[2] The international community has failed to agree on how to extend the 1992 United Nations Framework Convention on Climate Change (UNFCCC) at Rio de Janeiro and the Kyoto Protocol of December 1992.[3] The saga continues as I write in May 2011: The fifteenth Conference of the Parties, at Copenhagen, failed to reach agreement about quantitative emission reduction targets, with accompanying transparent "measurement, reporting, and verification" requirements.[4]

My purpose in this chapter is to offer the reader a framework for thinking about climate change. After a brief discussion of the science, I explain the restricted choices that face leaders of developed and developing countries and different ways to realize emission reductions. I discuss carbon capture and sequestration as an example of the difficulties of developing a significant carbon-avoiding technology. I conclude with an analysis of recent U.S. climate policy and argue that it is uncertain whether the policies under consideration would be effective, even if adopted.

The State of the Science

This is not the place, and I am not the expert, to give a convincing and accurate presentation on the science of climate change. The international scientific community of climate experts has been studying this problem for almost two decades through the mechanism of the Intergovernmental Panel on Climate Change (IPCC). The IPCC is dedicated to technical and socioeconomic assessment of information relevant to the understanding of the risks of human-induced climate change.[5] The important point is that there is a strong scientific consensus among qualified experts that continued emissions of greenhouse gases from human activity will cause climate change adverse to the global ecology and human welfare.

The IPCC assessments are based to a large extent on global climate models that are validated by the model's ability to reproduce past climate behavior. The IPCC assessments recognize that model predictions are uncertain (for example, due to the difficulty of modeling ocean interactions and the complexity of feedback effects caused by clouds) and therefore present a range of expected future mean global temperature increases, ΔT. It is generally accepted that if mean GHG atmospheric concentrations are kept below 550 parts per million, ΔT will remain below 2°C, a level that is considered acceptable.

I believe the uncertainty in the model predictions increases as one passes through the causal chain between emissions and impact on human activity:

GHG emissions → atmospheric concentrations → global temperature increase → climate change → impact on ecology and human welfare.

Accordingly, it makes good sense to keep an open mind about the state of the science and carefully consider objections raised by some to the conventional (and, I believe, accurate) IPCC analysis. One wants to be sure that experts do not overstate the case, and that new information, which possibly does not agree with the conventional analysis, is not ignored. Unfortunately the climate debate is often noisy and unproductive, because it is between passionate experts and critics who seem to question the science because it has adverse implications for their private interests or values. I believe the U.S. administration would be wise to sponsor, from time to time, a "competitive" scientific analysis that challenges the conventional wisdom to examine the assumptions, methods, and uncertainties in the IPCC studies.

What Are the Choices?

If the path to avoiding climate change involves global emission reductions, then all countries, especially countries with large or rapidly growing economies, must accept limitations on future GHG emissions. Nations have been negotiating how this might be done at fifteen Conferences of the Parties (COPs) established under the U.N. Framework Convention on Climate Change (UNFCCC), beginning in Berlin in 1995 and most recently in Copenhagen and Cancun.[6]

There have been two stumbling blocks to progress. The first is that during the eight-year George W. Bush presidency, the United States did adopt a climate policy. While the Bush administration generally accepted that global warming was real, it was ambivalent about the urgency of the problem and placed emphasis on developing new technology. The second, and more serious, stumbling block is the difference between the developing world and the developed world over who should bear the cost of emission reductions. I referred to this difference in the Introduction in terms of equity, but in this chapter I present Yoichi Kaya's powerful bit of analysis, which helps illustrate the problem facing different nations—such as the United States, as a developed country, and China, as a developing country.[7]

Kaya drew attention to an accounting identity that must be satisfied between different periods for the variables of economic activity (Y = GDP), energy use (E = energy), and GHG emissions (C = tonnes). The identity states that in any selected time period the fractional change in carbon emissions must satisfy the condition[8]

$$\frac{\delta C}{C} = \frac{\delta Y}{Y} + \frac{\delta (C/Y)}{(C/Y)}$$

$$\begin{pmatrix} \% \text{ change} \\ \text{in carbon} \\ \text{emissions} \end{pmatrix} = \begin{pmatrix} \% \text{ change} \\ \text{in GDP} \end{pmatrix} + \begin{pmatrix} \% \text{ change} \\ \text{in carbon} \\ \text{per unit GDP} \end{pmatrix}$$

The equation states that the percentage increase in carbon emissions between any selected initial and final period must equal the percentage increase in GDP between these two periods plus the percentage change in carbon per unit GDP (C/Y).

An expanded version of the Kaya relation is that the percentage increase in carbon emissions between any selected initial period and final period *must* equal the sum of the percentage increase in carbon intensity—that is, carbon per unit energy (C/E), plus the percentage change in energy per unit GDP (E/Y), plus the percentage change in GDP per capita (Y/P), plus the percentage population change:

$$\frac{\delta C}{C} = \frac{\delta (C/E)}{(C/E)} + \frac{\delta (E/Y)}{(E/Y)} + \frac{\delta (Y/P)}{(Y/P)} + \frac{\delta P}{P}$$

$$\begin{pmatrix} \% \text{ change} \\ \text{in carbon} \\ \text{emissions} \end{pmatrix} = \begin{pmatrix} \% \text{ change} \\ \text{in carbon} \\ \text{intensity} \end{pmatrix} + \begin{pmatrix} \% \text{ change} \\ \text{in energy} \\ \text{intensity} \end{pmatrix} + \begin{pmatrix} \% \text{ change} \\ \text{in GDP} \\ \text{per capita} \end{pmatrix} + \begin{pmatrix} \% \text{ change} \\ \text{in} \\ \text{population} \end{pmatrix}$$

Each country must consider how the policy it proposes to adopt will satisfy the identity.

In the 2010 *IEO* there is an official estimate of how different countries will satisfy the Kaya identity.[9] Table 2.1 presents the values given for the United States, China, and the world for the period 2007 to 2035.

There are several interesting points made by this table. First, the EIA is projecting an increase in GHG emission during the period 2007 to 2035 for the world and for the United States. Second, the difference in the challenge facing the rapidly growing emerging economies, exemplified in the table by China, and the large developed economies, exemplified in the table by the United States, is the expected difference in economic growth. Even if the emerging economies, such as China, are able to maintain an average reduction in GHG per unit GDP of $-2.8\% = \delta(C/Y)/(C/Y)$ greater than the expected reduction of the United States of $-2.1\% = \delta(C/Y)/(C/Y)$, their anticipated increase in carbon emissions will be at an annual rate of $+3\% = \delta C/C$. My expectation is that the large emerging economies like China, India, Brazil, Indonesia, Mexico, and South Africa will not be willing to sacrifice growth in order to stabilize emissions.

The third interesting point in the table is that U.S. population growth, largely due to immigration, is equal to the world average and larger than China's anticipated population growth rate.

Table 2.1. Kaya identity, average annual percentage change, 2007–2035

	$\dfrac{\delta C}{C}$	$\dfrac{\delta(C/E)}{(C/E)}$	$\dfrac{\delta(E/Y)}{(E/Y)}$	$\dfrac{\delta(Y/P)}{(Y/P)}$	$\dfrac{\delta P}{P}$
United States	0.3	−0.2	−1.9	1.5	0.9
China	3.0	−0.3	−2.5	5.5	0.3
World	1.3	−0.2	−1.7	2.3	0.9

These EIA projections contrast sharply to the expressed U.S. objective of reducing GHG emissions to 80% below 1990 levels by midcentury. This corresponds to a reduction from 2010 of about 85%, (or an annual average percentage reduction of $-4.7\% = \delta C/C$ over the period). If the United States experiences an average growth rate of only $2\% = \delta Y/Y$ per annum during this period, the United States would need to reduce carbon intensity by an average of $-6.7\% = \delta(C/Y)/(C/Y)$ per year over the period to satisfy the Kaya identity and achieve the target reduction.

The 6.7% reduction *must be* composed of the sum of reductions in energy intensity $\delta(E/Y)/(E/Y)$ and carbon-energy intensity $\delta(C/E)/(C/E)$. For the past several decades, the United States has experienced a 1.6% decline[10] in energy intensity per annum, $-1.6\% = \delta(E/Y)/(E/Y)$. (See Figure 2.1.)

Accordingly, the United States must reduce the carbon-energy intensity by 3.5–5.1% $= -\delta(C/E)$, a remarkably ambitious goal, the implication of which is discussed in the next section.

The House and Senate climate bills both contained a short-term goal of reducing GHG emission by 17–20% from 2005 emission levels by 2020. This corresponds to an annual carbon

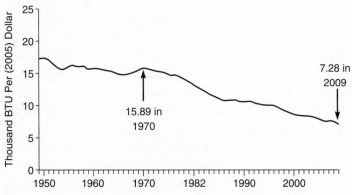

Figure 2.1. Energy consumption per real dollar of GDP, 1949–2008.

reduction of $-2.2\% = \delta C/C$. Assuming an economic growth of $2\% = \delta Y/Y$, the net carbon-energy intensity reduction must be $-4.4\% = \delta(C/Y)/(C/Y)$. With the historic decline trend in energy intensity, $-1.6\% = \delta(E/Y)/(E/Y)$, the required reduction in carbon-energy intensity must be a hefty $-2.8\% = \delta(C/E)/(C/E)$ per year.

Comparison of the 2010–2020 carbon reduction of $-2.2\% = \delta C/C$ per year, with the 2010–2050 carbon reduction, $-4.7 = \delta C/C$ per year, illustrates that U.S. climate legislation is "back-loaded" to delay the pain—for instance, even deeper carbon reductions in the period 2020–2050 would be required to make the target, $-5.6\% = \delta C/C$ per year.

China has a different challenge. In November 2009, the China State Council announced that China would reduce carbon dioxide emissions intensity by 40–45% from 2005 levels by 2020.[11] This would correspond to an annual reduction of carbon per unit output, $\delta(C/Y)/(C/Y)$, between -3.4% and -3.9%, greater than the -2.8% projected by the EIA. If one expects China's economy to enjoy an annual growth rate of about 6%, then we may conclude that China's carbon emissions, $\delta C/C$, will grow at an annual rate between 2.6% and 2.1% for the time period 2010–2020.

China and the United States each produce roughly 20% of world GHG emissions, so for at least the next decade, the result is that the emissions of China and the United States will increase in the range of 2.4% to 3.3% per annum. Over the longer term, if the United States realizes its optimistic performance assumption of reducing carbon emissions by an average of 4.7% per annum, and China maintains its economic growth and its ambitious annual carbon intensity reductions, so that its carbon emissions decrease by about 4.5%, the net emission growth of China and the United States will gradually increase, reflecting China's larger economic growth rate. The implication is that the world will blow through the level of 550 parts per million of GHGs well before the end of the century—leading me, along with many other observers, to believe that sooner rather than later other approaches (such as adaption and active geo-engineering) than emission reductions to avoid climate change will need to be seriously pursued.

The Kaya relation helpfully identifies the restricted range of choices that a country faces in achieving carbon reduction targets and the significant differences in choices likely to be made by different countries, such as the United States and China.

Effects of International Trade

Just as the domestic debate about climate change ignores international trade, so does the Kaya identity. The Kaya identity is expressed in terms of national income (Y), domestic energy use (E), and domestic carbon emissions (C). Although it is certainly possible to generalize the identity to take into account the effects of international trade, this is not customarily done.

The effects of trade are important. Consider two possible ways that the economy might adjust in order to achieve an improvement in energy efficiency (lower E/Y). First, assume there is no trade, and that the country achieves its emission reduction target by replacing domestic energy intensive facilities by more energy efficient industries that meet consumer demand, although, perhaps, at higher cost. If the more energy efficient industries substitute for the output of the energy intensive industries, a net global reduction in emissions results.

But if trade is possible, the country may import the products to replace the lost output of the energy intensive industries. To the extent that the exporting facilities are less efficient than the domestic energy efficient facilities posited above, there will, to some extent, be less net global emission reductions—the reduction in emissions from inefficient domestic facilities is offset by imports coming from increased production at inefficient international facilities.[12] Thus local improvement in efficiency does not lead to a corresponding global reduction in emissions.

Absent global compliance, energy intensive industries will be disadvantaged in countries with stringent constraints on emissions. The inefficient industries will move to countries with more lenient emission controls and produce products and services for export to markets where emission constraints are imposed. When asymmetric emission constraints exist in different countries, the result of international trade may be that emissions are effectively exported, to some degree. In the country with controls, energy intensive industries become less competitive, which suggests imposing tariffs on the energy intensive products from countries without controls that meet the domestic demand displaced by the energy efficiency adjustment.[13]

Different Ways to Realize Emission Reductions

The Kaya identity states that different desired reductions in carbon emissions must be met by reducing energy intensity (E/Y) and carbon-energy intensity (C/E). But there are many ways that this might occur.

Improving energy efficiency is the most cost-effective choice. Improvements in energy efficiency are accomplished by (1) regulatory mandate, such as requiring automobiles to satisfy a certain level of fuel economy or mandating efficiency standards for appliances; or (2) raising energy prices as a market incentive for energy users to use less energy. Of course, high energy prices shift the basket of products and services produced by the economy to less energy intensive items, but, as discussed above, this may amount to moving energy intensive industries, such as cement, steel, and chemicals, offshore, with no net global reduction in emissions.

Although the United States has enjoyed decades of consistent reduction in energy intensity, much more improvement is possible. Some witch doctors prescribe the happy, but not demonstrable, message that the U.S. economy can achieve improved energy intensity at lower cost. It is surely true that if best practice spread to throughout an entire sector, its average energy utilization costs would decline. However, the puzzling question is: Why does the private sector not pursue opportunities for negative cost efficiency investments if they exist? Most energy efficiency improvement occurs because of regulatory mandates or economic incentives brought about by higher prices. So although efficiency improvement is the first answer, it is not the only answer.

The second possibility is to reduce carbon-energy intensity. In practice, in the near term, this option means fuel switching from coal and oil to natural gas and nuclear. In the longer term it means a shift to renewable sources of energy—solar, wind, geothermal, biomass. These longer-term alternatives will grow in importance over time as the cost of conventional sources of energy rise, and as technology advances.

The key issue is coal. Coal accounts for about 50% of U.S. electricity generation and 35% of U.S. GHG emissions. In China, where electricity growth is projected to be about 4.5% per year (compared to approximately 1% in the United States and Europe), in 2006 coal amounted for 79% of electricity and 77% of China's GHG emissions.[14] Because coal costs about $1–2 per million BTUs, compared to natural gas at $6–10 per million BTUs, it is unrealistic to imagine that the United States and China will abandon its use. The EIA estimates that emissions from coal will grow between 2006 and 2030 at an average annual rate of 0.7% in the United States and 3.1% in China. It is estimated that coal will maintain its market share of about 26% for all primary energy through 2035.[15] The challenge is to find a cost-effective way to burn coal and not emit carbon dioxide, the principal product of combustion, into the atmosphere.

How Might the Emission Reductions Be Realized?

Legislative proposals contain a mix of the measures described above, intended to achieve a desired level of emissions reduction. Several governmental and nongovernmental organizations model the possible outcome of alternative climate legislation proposals.[16] These models differ in detail and complexity, but all share a dependency on assumptions about critical market factors: demand and supply response to price changes, costs of various technologies that are as yet unproven, and rates of penetration of new technology. Any number of models could be chosen to illustrate the range of uncertainties in outcome covered by a reasonable range of assumptions. I choose to consider the predictions of the MIT Emission Prediction and Policy Analysis (EPPA) model derived from their analysis of the Waxman-Markey climate bill.[17]

The MIT EPPA model is designed to estimate the future of the U.S. or world energy economy based on a precise set of input values for the economic activity and a specification of the adopted climate policy. The EPPA models energy markets where industrial sectors make choices and select fuels based on economics; fuel prices are determined endogenously in the model, based on parameter input values that characterize supply and demand curves.

Figure 2.2 (on page 42) gives the results of the MIT EPPA model to midcentury for five different scenarios that give relative advantage (in terms of technical readiness and performance) to different technologies:[18]

Scenario 1 is business-as-usual, no climate change legislation is enacted, and the results show the resulting baseline energy use and GHG emission at midcentury.

Scenario 2 shows a reference response to meet the 80% GHG reduction target by 2050. The CO_2 equivalent charge is $229 per tonne in 2050 in 2005 dollars.

Scenario 3 assumes a relative advantage for nuclear power. Here the relative advantage includes the assumption that nuclear power becomes publicly acceptable and that it is economically attractive relative to alternatives.

Scenario 4 assumes that carbon capture and sequestration for coal is available at a relatively attractive cost.

Scenario 5 assumes that (nonbiomass) solar technologies, such as wind and photovoltaics, are relatively more attractive performers.

Of course, in each of the scenarios the preferred fuel makes the largest contribution, but there is significant variation among the different scenarios. For example, carbon capture and sequestration (CCS) provides 21% of the energy in its preferred scenario 4, but 0% in scenario 5.

Third, in three of the four alternative scenarios, the biggest contributor to reducing carbon emissions is the energy demand reduction from the carbon constraint.

The analysis shows that there are several ways the economy might adjust to a mandated emission reduction of 80% by midcentury, but of course the model analysis does not indicate the likelihood of each scenario, which reflects the uncertainty in the input parameter assumptions. The most important uncertainties are the readiness of the various technologies, their cost, and the pace at which the technologies will be adopted.

Carbon Capture and Sequestration and Government Action

I discuss one of the technology choices—carbon capture and sequestration—in order to illustrate the scale and the roles and

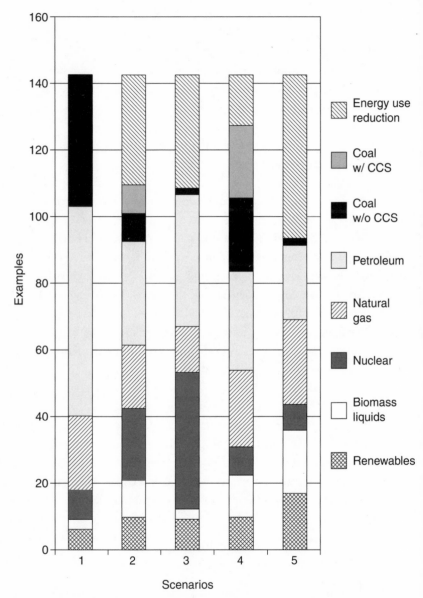

Figure 2.2. Primary energy share, 2050.

responsibilities of industry and government in creating a serious technology option for reducing carbon emissions. The analysis presented in this section is taken from the 2007 interdisciplinary MIT study, *The Future of Coal.*[19]

Figure 2.3 shows a modern 500 MWe (megawatts of electricity) ultra-supercritical coal-fired power plant that produces electricity by burning pulverized coal in a boiler and using the steam to produce electricity in a turbine. This plant operates at high temperature to achieve an efficiency of about 43%. Coal types vary widely in energy content per tonne (for example, bituminous coal from the Midwest has higher sulfur but more energy per pound than coal from the Powder River Basin in Wyoming). A supercritical 500 MWe pulverized coal plant will consume about 5,000 metric tons per day of coal and emit about 18,000 tonnes per day of CO_2, while producing 12 million kilowatt-hours of electricity per day (kWe-hr per day), when burning Illinois #6 bituminous coal.

In 2005 the United States emitted about 6 billion tonnes of CO_2 equivalent with coal-fired electricity generation accounting for about one-third of this total.[20] This means that to reduce CO_2 emissions by 1 billion tonnes (about 15% of the total), 55% of all coal-fired power plants would need to have the capability to capture 90% of the CO_2 emitted.

How are coal plants modified to reduce CO_2 emissions? For new plants, there is a range of technology choices. In the United States the preference is to use integrated gasification combined cycle (IGCC) plants, which first gassify coal with oxygen and then add steam to produce "syngas," a mixture of carbon monoxide and hydrogen. The syngas is sent to a water-gas shift reactor unit where steam is added and reacts with the syngas to produce a stream of hydrogen and CO_2. The CO_2 is captured in a capture unit, compressed, and transported by pipeline to a

Pulverized Coal Plant without CCS

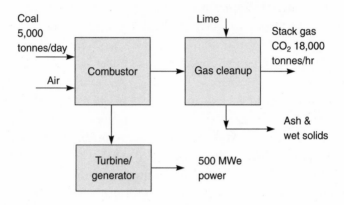

IGCC Coal Plant with CCS

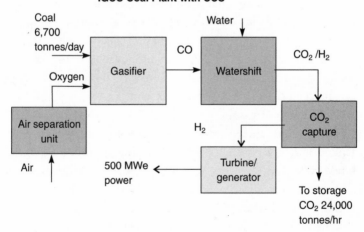

Figure 2.3. Coal plant.

storage site. The hydrogen is sent to a combustion turbine to produce electricity.

There are three unpleasant realities with the CO_2 capture scheme. First, the process is less efficient, requiring more coal per unit of electrical output; the anticipated efficiency of such plants (there are none operating in the world today) is expected to be about 32%. This means that the IGCC plant with CO_2 capture consumes about 6,720 tons of coal per day and emits 24,200 tons of CO_2 per day to produce 500 MWe. Second, the capital cost of an IGCC plant with capture is higher than a supercritical pulverized coal plant because of the lower efficiency. Third, more coal is consumed and thus fuel costs will be higher.

Europe prefers an alternative approach for new plants, Oxy-fired pulverized coal combustion. In this scheme, pulverized coal is burned in oxygen rather than air. The nitrogen in the air is removed before combustion so that the combustion gas is mainly CO_2 and therefore easier to capture.[21] There is no reason to favor either technology at this time; the ultimate difference will depend on coal type and project specifics.

Economics for new plants. The MIT coal study estimated the cost difference for a 500 MWe plant between pulverized coal combustion without capture and IGCC with 90% capture (see Table 2.2).[22] Evidently, because there are no such plants in operation and because operating and financial assumptions differ, there is a wide variation in the cost estimates of coal-fired generation with or without CO_2 capture. The estimates from the MIT coal study presented below[23] do not include the cost of transportation, storage, and monitoring, about $5 to $10 per tonne, and may be compared to DOE estimates.[24] When sequestration costs are added, a rough rule of thumb is that the cost of generation for CCS will be 50% higher at the point of production—that is, before distribution to customers—than the cost of generation without carbon capture. Since generation cost is approximately one-half of cost to the end-use, because of distribution charges, it is reasonable to suppose that removing carbon emission from the electricity-generating sector will result in about a 25% increase in total cost to the end user.

The cost of $32 per tonne of CO_2 avoided is high, but less than the cost of other carbon abatement alternatives, such as replacing motor gasoline with biofuels. Developed nations may be able to absorb the increase in electricity from CCS, but developing countries will be loath to do so.

Retrofit. A supercritical pulverized coal plant emits about 6 million tonnes of CO_2 per year. Construction of 166 new 500 MWe plants with CCS is required to displace one billion tonnes of CO_2 annually, which is 15% of the U.S. total. The incremental investment cost for 90% CO_2 capture is about $400 million per plant. Because of this high cost and the long lifetime of coal plants (forty years or more), it is relevant to inquire about the prospects of retrofitting conventional air-driven pulverized coal plants for CO_2 capture. The unfortunate conclusion is that retrofit of existing plants is neither technically easy nor cheap.[25]

Table 2.2. Cost differences for a 500 MWe plant

	Supercritical pulverized coal w/o capture	IGCC with 90% CO_2 capture
Total plant cost $/kWe	$1,913	$2,719
COE ¢/kWe-h	6.26	8.59
CO_2 emitted g/kWe-h	830	102
Cost of CO_2 avoided $/tonne	NA	31.9

Sequestration. Once the CO_2 is captured, it must be sequestered from the atmosphere. There are many different schemes for how this might be done,[26] as illustrated in Figure 2.4, from the IPCC 2005 report on CO_2 capture and storage.[27]

It is generally agreed that the best prospects for CO_2 storage lie in deep saline aquifers, and that there may be some limited, early opportunities in enhanced oil recovery (EOR). The requirements for a properly functioning CO_2 sequestration project are these:[28.]

- Site selection of adequate storage capacity (millions of tonnes of CO_2).
- Safe procedures to inject significant quantities of CO_2 (millions of tonnes per year) into the aquifer.
- In-place sensors to measure, monitor, and verify the integrity of the storage.
- Determination of liability responsibility for the storage site and possible leakage over long periods of time.
- A publicly acceptable environmental framework that establishes regulations, inspection, and enforcement of the sequestration project operations. Environmental groups believe that large-scale demonstrations are necessary, but they are understandably concerned about the efficacy of the regulatory framework.[29]

An integrated CCS system consists of capture at the point of CO_2 formation, pressurization and transport by pipeline of the CO_2, and injection of the CO_2 at the storage site. The scale is enormous, with the mass flow of CO_2 by pipeline potentially greatly exceeding the mass flow of natural gas in the current pipeline system—although the extent of the pipeline system might be less, depending on the location of the sequestration sites relative to the plants generating CO_2.

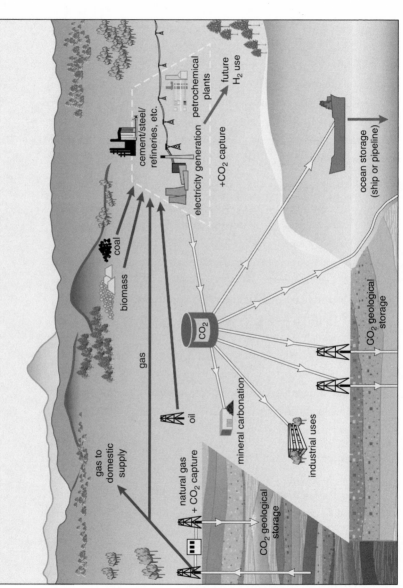

Figure 2.4. Schematic diagram of possible CCS systems. *Source:* Figure TS-1 from page 20 of the technical

The following labels appear in the figure:

gas to domestic supply

natural gas + CO$_2$ capture

CO$_2$ geological storage

mineral carbonation

industrial uses

oil

gas

biomass

coal

cement/steel/refineries, etc.

petrochemical plants

future H$_2$ use

electricity generation

+CO$_2$ capture

CO$_2$

CO$_2$ geological storage

ocean storage (ship or pipeline)

MIT maintains an informative website with a database of operating and planned carbon capture sequestration projects around the world.[30] There are no CCS coal plants operating at scale (500 MWe or greater) in the world, although some projects are planned, and several small-scale projects (less than 100 MWe) are in operation. There are three large (1 million tonnes per year or greater) sequestration projects in operation: the Statoil Sleipner project in Norway, the Weyburn project in Canada, and the BP project in In Salah, Algeria. In addition, the Chevron-Texaco project in Gorgon, Australia, will begin operation in 2014. All of these projects sequester CO_2 separated from gas production activities; monitoring is limited to seismic 4D surveys, and there is no explicit environmental framework set for the activity. In sum, these and many other smaller operating or planned sequestration projects around the world yield useful technical information, but the pace is insufficient to establish CCS as a credible technology option within the next decade. So policymakers and legislators may fantasize that CCS technology will be proven and available for general deployment around the world by 2020, but this is most unlikely to be true.

CCS: What Should Governments Do?

The International Energy Agency (IEA) has issued an "energy technology analysis" of CO_2 capture and storage that presents ambitious scenarios for deployment of CCS.[31] One of the scenarios projects that a \$50 per tonne CO_2 emission charge would result in 5.1 billion tonnes of annual CO_2 capture and storage, with 18% of total electricity generation from plants equipped with CCS, by midcentury. The study states that CCS is an important cost-effective option for reducing GHG emissions.

The government's role is not to choose which GHG emission technology options are "winners" and should be deployed.

The government does not have the knowledge or the expertise to do so, the performance and cost of the various options is quite uncertain, and the evolution of energy markets is unknown. The proper government role is to establish the technical performance, economics, and environmental effects of each candidate technology, so that the private sector can make informed investment decisions about what is most cost-effective.

The set of actions that the government needs to take to establish CCS as a credible technology option have been clear for some time:

- The government should fund the design, construction, and ten-year operation of three to five carbon sequestration projects at a scale of one million tonnes of CO_2 per year.
- In parallel, the government should establish a regulatory framework for CO_2 sequestration, conduct R&D to resolve technical issues related to sequestration, and explore a system for the near-real-time monitoring of CO_2 storage. The cost of such a program would be about $400 million per year.
- The government should provide financial assistance to industry to demonstrate three to six integrated first of a kind clean coal systems with CCS for a range of technologies: IGCC coal power plants, Oxy-fuel pulverized coal power plants, capture retrofit for existing coal-fired power plants, and coal-to-liquids or coal-to-gas plants with carbon capture. The cost of each such integrated plant could be as high as $2 billion and the government cost share perhaps 25%.

CCS: What the United States Has Been Doing

The DOE budget for carbon sequestration was relatively modest, reaching a level of $140 million in FY2009.[32] The DOE program has focused on R&D, support for seven regional partnerships in the United States, and participation in several international collaborations.[33] For the first years of its existence, the regional partnerships program was relatively small and focused on engaging regional groups in discussing problems of "characterization and validation."[34] Many small projects offered an opportunity to involve stakeholders. Beginning in 2008, the regional partnerships entered a "development" phase undertaking credible projects. For example, the Midwest Geological Sequestration Consortium (MGSC) envisions a seven-year $84 million project to store CO_2 collected from the Archer Daniels Midland Decatur, Illinois, ethanol plant, through a 6,500-foot injection well, into a porous saline rock formation.[35] One million tonnes will be injected over the lifetime of the project. But as late as March 2010, the federal government had yet to launch an interagency effort, involving at a minimum the Environmental Protection Agency and DOE, to define a comprehensive regulatory framework for CO_2 sequestration. The result is a proliferation of divergent state rules that will prove difficult to reconcile.[36] The level of activity and resources falls far short of what is needed to establish CCS as a credible technology option, and far short of what has been recommended by the MIT Future of Coal Study[37] and others.[38]

The 2009 American Recovery and Reinvestment Act stimulus package has provided a large pulse of funds for CCS. In June 2009, the DOE announced an agreement to provide up to $1 billion for the FutureGen project for the construction of the first commercial-scale IGCC plant with CCS in Mattoon County, Illinois. But in another illustration of the lack of durability in government policy, in August 2010 the DOE announced FutureGen2.0 with an altered focus to repowering an existing coal plant with CCS, the project still to be located in Illinois.[39] In addition, in December 2009 the DOE announced $979 million of government support (with funding from the stimulus package) for three CCS projects sponsored by American Electric Power Company, Southern Company, and Summit Texas Clean Energy LLC.[40]

This is big money, but it is still not on the right track. All the projects are for integrated power generation, capture, and sequestration. The priority should be for sequestration projects that are cheaper and a prerequisite for integrated CCS. The projects are proceeding to be designed in the absence of a national regulatory framework for sequestration. The DOE has limited capability to manage such complex integrated projects, as will be discussed further in Chapter 4. It is not clear whether funding will be available, once the stimulus package expires, to pay for the inevitable overruns or to support the collection, analysis, and dissemination of data to other potential investors, which is the key justification for government support of demonstration projects of this type.

The analysis in this chapter of CCS, an especially important and complicated technology option, shows that the U.S. government has not yet arrived at an effective way to manage complicated RD&D programs. The failure is not a matter that can be solved by simply spending more money, although additional resources will certainly be needed. The U.S. government and the DOE in particular must do a better job of managing energy innovation, as I will describe in Chapter 5. And the executive and legislative branches must recognize that energy innovation takes a long time and there must be agreement on long-range technology plans.

Climate Change Legislation in President Obama's First Two Years

President Obama came into office in 2008 making it clear that climate change legislation that would reduce GHG emissions was a priority for his administration. President Obama favored a cap-and-trade system as a mechanism for emissions reduction.[41] (A tax on emissions, which I happen to prefer, is an alternative mechanism that could be structured to give an equivalent emission outcome, but that is not the point here.) The adoption by the United States of an emission reduction target at this level, if realized, would be an important step toward solving the dilemma regarding global climate changes. The second, and arguably more difficult, step is to ensure that the large rapidly growing developing countries agree to reduce their emissions.

President Obama pledged as late as 2010 "to find the votes in coming months" for legislation that includes 80% GHG emission reductions by the mid-twenty-first century.[42] Imagine that you are the president, or one of his senior policy advisors, or a legislator. If you are responsible, then you will want to have reasonable confidence that the policy mechanism authorized by the legislation will achieve the emission reduction goal. The only means to gain this confidence is to examine the results of model projections. In reality, as we have seen, the range of outcomes is tremendously uncertain; achieving targeted reduction depends upon the assumptions about technical performance, cost, and market response proving accurate. Curiously, many policymakers and legislators seem to believe that once Congress passes legislation, the market will adjust to adopt less carbon intensive processes and technologies without any additional enabling measures or reference to economic conditions.

The Obama administration turned to Congress to craft climate legislation—the House led with passage of the Waxman-Markey bill, and the Senate spent a great deal of 2009 and 2010 debating the Kerry-Boxer bill and then a closely related Kerry-Lieberman-Graham variation. As all these pieces of legislation passed through the legislative process, and compromises were made (especially because the administration was not active defending its proposal) that sharply limited the prospect that the goals set could be achieved. For example, the time and schedule of emission caps were lengthened, and special provisions were extended that limited coverage of greenhouse gas emissions. Emission allowances were allocated to many emitters for free for a period of time, rather than auctioned, thus dampening the desired effects of a carbon charge. Significant offset credit provisions to meet future emission obligations (up to 30%), notably for the domestic agricultural sector international abatement projects, were added to attract support of influential stakeholders because offset lowered the near-term cost of compliance.[43] Indeed, more than one business leader admitted to me that their support for the climate change legislation was based on the result of company analysis that concluded that passage of the legislation would have no effect on company plans for at least fifteen years. While genuine offsets are an important aspect of a comprehensive emission control policy,[44] the specific provisions included in the legislation seemed motivated by attracting political support, thus raising serious questions about the process for setting baselines and monitoring compliance.

The net result is that the cap-and-trade legislation of both the House and the Senate effectively set extremely low emission charges, as illustrated by the "back loading" discussed in the section on the Kaya identity above. I doubt that the deep reductions envisioned in the legislation could have been achieved in the time period specified—that is, 80% reductions by 2050—because the measures were too weak to achieve the demand reduction, fuel switching, and economic incentives to introduce new technologies such as CCS, which, in any event, were based on overly optimistic projections of technology readiness of carbon-free energy options. The models predict that 80% emission reduction by midcentury is possible only if there is a significant carbon charge *and* a remarkably fortunate technical and cost experience with new technology.

The 111th Congress did not pass climate change legislation sought by President Obama. Unquestionably, the main reason for this failure was the financial crisis of 2009, which turned attention to restoring employment and growth and away from other initiatives that, no matter how meritorious, were seen as adversely affecting U.S. economic competitiveness. Some also fault the Obama administration's legislative tactics.[45]

I, for one, do not regret the failure for the legislation to pass. The legislation did not set a credible price signal or other measures that gave any chance of achieving the ambitious GHG emission reduction goals. This is an example of the danger, discussed in Chapter 1, of setting unrealistic goals. Target reductions, especially when based on unrealistically optimistic technology assumptions, will result in sharply higher prices for emission allowances, which in turn means higher costs for producers and consumers as well as difficult market adjustments. The inevitable result will be that political pressure will crater the unrealistic carbon control policy.

Any hope for passage of climate legislation based on a cap-and-trade emissions tax vanished with the midterm 2010 election. This presents a dilemma for the Obama administration. The administration correctly remains convinced that increasing domestic and global emissions inevitably will result in major climate disruption, but it removes the policy tool—emission charges—that most directly and efficiently will reduce the risk of climate change.

It is sad but understandable—because "politics is the art of the possible"—that the Obama administration must resort to reliance on second-best measures to achieve emission reductions. First, the administration allocated a significant amount of stimulus funds ($33 billion) to energy projects that were judged to advance carbon-free technologies.[46] This clearly is a short-term measure. Second, administration rhetoric shifted from "the need to avoid the risks of climate change" to the potential of energy legislation to create new "green jobs" and reduce dependence on imported oil. The validity of this new proposition is dubious and certainly unsupported by analysis. If the government funds energy programs, it is sure that additional jobs are created—that is a necessary consequence of spending more money on activities that employ people. But the administration has not convincingly demonstrated that "green energy programs" create more *net* U.S. jobs than are created by other possible programs. Moreover, creating jobs, if they are for economically uncompetitive activities, is not sustainable in the long run. Third, the Obama administration has stated that it will pursue emission reductions using the authority of the Clean Air Act, which the Supreme Court has ruled extends to greenhouse gas emissions.[47] But this risks abandoning market-based mechanisms that are most efficient and powerful (as demonstrated in the history of the cap-and-trade clean air program) for mandates such as renewable electricity standards or renewable fuel standards (a low carbon standard would be preferable) that are manifestly less economically efficient than a cap-and-trade system.

Unfortunately there has also been no progress in reaching an international agreement on global emission reductions between developed and developing countries. As mentioned in Chapter 1, the high expectations set for the 2009 Copenhagen conference were not met, and no significant results were expected or realized from the 2010 Cancun conference. In sum, during Obama's first term, U.S. diplomatic climate initiatives have achieved little, despite genuine effort.

President Obama's retreat from his energy policy was skillfully accomplished given the political realities in his energy speech on March 30, 2011.[48] His message shifted from the politically unpopular danger of climate change and the need for a cap-and-trade system, to the politically more popular need to reduce oil import dependence. The centerpiece of the new policy is a clean energy standard requiring 80% of U.S. electricity generation to come from low carbon sources by 2035: natural gas, nuclear, and renewables (mostly hydropower and wind), although crucial details of the standard have not been presented. Greater domestic oil production, increased biofuel use, and natural gas and electric vehicles are expected to replace imported oil. The president mentioned many past presidential speeches had noted the urgency of action addressing energy issues but that little had come to pass. Yet he proceeded to set similar goals, in similar ways as in the past.

The administration issued a "Blueprint for a Secure Energy Future," to support the new policy.[49] But, this blueprint is not a comprehensive plan that addresses all-important aspects of energy; nor does it contain metrics on which to judge the progress of the many disconnected initiatives that are proposed. No analysis is presented of how the different aspects of the policy relate to each other and adjust to possible changes in the domestic and international economy. The effect of the proposed measures on energy prices is not mentioned—importantly the indirect effect on prices to consumers of requiring utilities to meet a clean energy standard by producing or purchasing clean electricity generation. This blueprint is yet another example of cobbling together many well-intentioned initiatives without priority as to their relative costs and benefits. Rather than providing a robust policy framework for the future, it reminds the public of how little progress is being made. The one positive feature of the blueprint is reaffirmation of the administration's commitment to continue investing in energy research, development, and demonstration.

Lesson Learned

Science shows that adverse global warming can be avoided if GHG emissions can be reduced. Reductions will occur through regulation or through economic response if a significant emissions charge is imposed (through either an emissions tax or a fancier cap-and-trade system). Developed and developing countries face different constraints in adjusting to emission reductions, and consequently different domestic policies should be expected. The implacable nature of the Kaya identity reveals how these constraints influence the positions adopted in international climate negotiations between developed and developing countries. It is far from clear how the United States and other countries will convince developing countries to adopt policies that will reduce growth in emissions, especially when high economic growth is expected. It is extremely unlikely that the United States will adopt a policy with a charge on emissions that is sufficient to realize 80% reductions in GHG emissions by 2050. Moreover, the readiness of technologies (such as CCS) required to achieve these reductions is very uncertain given the pace and nature of the U.S. government's development efforts.

The level of GHGs in the atmosphere in preindustrial times was 330 ppm in CO_2 equivalents. GHG concentration levels today are 470 ppm in CO_2 equivalents, causing an increase in global mean temperature of 0.8°C. If GHG concentrations level out at 550 ppm in CO_2 equivalents, there is an 80% chance that the temperature increase will exceed 2.0°C. Without stringent global emissions control, GHG concentrations will continue to increase and the chance of average global temperature increases between 4.0°C and 6.0°C by 2100 become significant. The EIA Kaya projections presented above indicate continued global increase in GHG emissions at least until 2035. All indications are that the 550 ppm level will be exceeded and the world will need to adjust to significant climate change.

The point is that the U.S. climate policy is not succeeding at either the domestic or the international level. Unrealistic goals and reduction targets that are based on optimistic hopes about technology readiness and effectiveness of compromised emission control polices have not had, and will not have, the desired effect of putting the country on a path to an effective climate policy. The unhappy conclusion is that policy analysis should now be devoted to adaptation as well as mitigation as a way to avoid the risks of climate change.[50]

3

ENERGY SECURITY

Energy security refers to the connection between energy markets and national security in production, transmission, and use of energy. The energy security landscape has two dimensions: one economic and the other political/military, with a contour that depends upon the particular energy source under consideration.

For coal the security challenge is to reduce risk of climate change and control the emissions of CO_2; here national economic interests intersect with global geopolitical issues. For nuclear power, the challenge is to reduce the risk of proliferation of nuclear weapons associated with the nuclear materials in the fuel cycle; here the political/military dimension dominates.

For oil and gas the economic energy security dimension involves the cost of market disruption that accompanies an interruption of supply. The political/military dimension involves the influence of import dependence on the foreign policy of the United States, its allies, and its adversaries. Import payments transfer significant dollars to oil-exporting countries that may not be in sympathy with the values or interests of the United States and its allies.

Oil and Gas Import Dependence and Energy Security[1]

Import dependence becomes a problem for the public and for political leaders when the amount of the commodity imported is a significant fraction of the total usage and the commodity is vital to the functioning of an economy, making reliability of supply critical. These conditions are not unique to energy, but the scale and ubiquity of energy attracts the most attention.

The principal economic concern is that an abrupt supply interruption is likely to cause a price shock that disrupts the economy. A price shock occurs because users of energy bid up the price of available supply since they cannot quickly adjust to using less energy or switch to other fuels. There are other economic consequences of ongoing dependence, including a large deficit item in the current balance of trade account, an adverse effect of the high cost of energy on competitiveness of energy intensive products, and the perceived net loss of jobs resulting from importing energy from abroad. Most economists believe that the best protection against supply interruption is to encourage diversity of supply and transparent, competitive energy markets.

In 1960 a group of producers organized OPEC, the Organization of Petroleum Exporting Countries, to control the supply and price of oil in the dollar-denominated world oil market. The experience of the 1973 and 1979 oil supply disruptions has made reliability of supply and avoidance of the price shocks accompanying supply disruption a priority concern for import dependent countries for the past three decades. Analysts have produced a considerable literature on the economic consequences of price or supply disruption for consuming countries due to intentional OPEC cartel action. However, concern has diminished over the years, because oil has reduced importance in the U.S. economy. For example, the sharp increase in real prices of oil in 2008 did not have a major effect on the economy compared to the interruptions of the 1970s.[2] Moreover, there have been few instances when OPEC has attempted directly to restrict supply in the past three decades. In general, OPEC's effort to control oil prices has had mixed results.

On the other hand, concern remains high about the impact of oil import dependence on the foreign policy of the United States and other consuming countries. The linkage of import dependence to foreign policy is not through OPEC's ability to directly impact world oil prices or create oil supply shocks. Rather, the linkage comes from the influence of import dependence on the political behavior of producing and consuming countries.

Producing countries attempt to control supply, and importing countries seek to secure supply through state-to-state agreements, each in order to improve their particular advantage. The geopolitical consequences of such supply arrangements cannot be interpreted solely in economic terms. Because import dependence means that a nation relies on foreign sources of supply for a commodity that is critical to the functioning of its economy, dependence inevitably influences the foreign policy of the importing country, and its reluctance to jeopardize supply will constrain, to some extent, the country's options in pursuing broad foreign policy objectives.

For the exporting country, its significant reserves and its production of oil and/or gas are important leverage for achieving its policy agenda. Differences in the policy agenda between consumer and supplier states will be quite pronounced for resources located in parts of the world, such as the Persian Gulf, that are unstable and have major resource holders who are not particularly friendly to the United States and its allies and/or their objectives.

Iran presents an excellent example. The fact that Iran offers about three million barrels of oil per day on the world oil market constrains the United States and several of its important allies that depend on oil imports, such as Germany, France, and Japan, from pressing Iran too strongly on issues such as its covert nuclear weapons program or its sponsorship of terrorism in Iraq, Afghanistan, Lebanon, and Gaza.

How the Geopolitical Aspects of Natural Gas and Oil Differ

The security aspects of oil are well known. Oil reserves are found principally in the politically unstable Persian Gulf (Iran, Iraq, Saudi Arabia, and Kuwait) and in countries that are not friendly to the United States (Venezuela, Russia). There has been no overt OPEC action to restrict supply since the 1970s. More recently the concern has turned to competition for access to supply, through state-to-state agreements between the industrialized OECD countries and the rapidly growing, large emerging markets, such as India and China.[3]

A portion of the oil revenues that flow to those major resource holders that are unfriendly to the United States are put to bad uses, such as financing terrorists or insurgent groups. For example, it is widely reported that Iran finances the terrorist activities of Hezbollah, and Venezuela provides funding to insurgent groups in Latin America. Clearly this activity presents security concerns for the United States and the countries that are affected by interference from hostile states.

The security aspects of natural gas are similar, but not identical, to those of oil. Natural gas imports play a smaller role in the economies of most importing countries compared to oil, mainly because it is less costly to transport liquid crude oil and petroleum products than natural gas. There are exceptions. For example, Singapore and Japan are highly dependent on imported natural gas, and Europe, understandably, shows increasing concern about its dependence on natural gas from Russia and North Africa. Natural gas is transported by pipeline over long distances, and the pressurization costs of transmission, as well as the need to finance the cost of these pipelines, encourages long-term contracts that dampen price volatility. On the other hand, because those who control the pipelines have the power to interrupt supply, the competition over the siting and operation of pipelines has both economic and political aspects.

Natural gas is an increasingly attractive fuel, though, because of its clean burning characteristics compared to oil or coal, and because of its price advantage, on an energy equivalent basis, compared to oil. Accordingly, the expectation is that natural gas will enjoy future growth in worldwide consumption and trade. Significant investments are being made to meet this future demand by bringing "stranded" gas to market. At distances between production and consumption of greater than 2,000 km offshore and 4,000 km onshore, it is more economical to transport natural gas either in liquid form, as liquefied natural gas (LNG), or perhaps as a liquid product, such as methanol. The trend is that natural gas will gradually become a global commodity with a single world market, just like oil, adjusted for transportation differences. Natural gas in Qatar and Nigeria, once considered stranded, is now being sold as LNG in world markets. Iran's natural gas remains stranded because of political factors.

As the natural gas market becomes more global, there will be a move to a world price for natural gas. At present there is a large disparity between natural energy equivalent natural gas prices in the different world markets—North America, Europe, and Asia—and prices for oil. In Asia the prices are close to energy parity, whereas in the United States natural gas sells well below oil on an energy equivalent basis. The reason for the disparity is that in Asia, and to a lesser extent in Europe, natural gas substitutes for oil use, whereas in the United States natural gas is a substitute for coal in the power sector. However, over time, as trade increases and the economic incentive leads to new technical opportunities for natural gas to substitute for oil—for instance, in the transportation sector—it is likely that the price of oil and gas will come closer to a global equivalence based on energy content.

Moreover, dramatic technical developments are expanding the economical recoverable natural gas resource base. Exploration and production are moving to deep water and offshore. The development of technologies to produce gas from unconventional sources—tight gas, coal bed methane, and, most importantly, gas from shales—means that it is unlikely that the United States, and eventually many other countries, will be dependent on imports of natural gas to any significant extent for some time. For at least the next decade, therefore, the principal concern will be with the security implications of oil import dependence.[4]

Security Implications of Global Trade in Oil

Several trends suggest that security concerns over oil import dependence will grow over the coming years. First, all experts project an inexorable increase in global oil demand, with especially rapid demand growth from the larger emerging economies, such as India and China.[5] Although the current global economic downturn provides a period of respite from increasing consumption and high prices, most experts believe that there will be a return to growth and therefore increasing demand for oil and gas.

Advances in technology will continue to improve the productivity of oil and gas exploration and production, and perhaps the efficiency of oil and gas use, but not sufficiently to offset a long-term trend to greater use and higher real prices. Accordingly, both producing and consuming countries should anticipate, at least for the next two decades, increased demand and the accompanying geopolitical implications, absent significant policy initiatives to reduce demand.

Second, increased production will be required from the Persian Gulf—Saudi Arabia, Kuwait, Iran, and Iraq—as well as from other countries that are either politically fragile or unfriendly to the United States and others, such as Nigeria, Ecuador, Venezuela, and Russia. This means that consuming countries must learn to work together to advance their interests with the major resource holders and to recognize the importance of maintaining political stability, especially in the Persian Gulf. One important step that would improve the leverage of consuming countries would be to admit China and India to the International Energy Agency.

Third, national oil companies (NOCs), such as Saudi ARAMCO and Brazil's Petrobras, are increasing their control of reserves and production compared to the large international oil companies (IOCs), such as Exxon-Mobil, BP, Total, and Chevron. The NOCs, many of which are very competent in managing oil exploration and production, serve the interests of their governments, which view their oil resources as a means to advance their political objectives as well as to obtain revenue. The result is an increasing trend toward state-to-state agreements between producers and consumers, especially with the new consumer countries. A noteworthy example is China's arrangements with Angola and Sudan. State-to-state agreements are an undesirable move away from open and transparent world oil markets toward the use of oil as an instrument of political influence. As mentioned above, Iran's 3 million barrel per day oil export constrains the European Union's willingness to take action against Iran's nuclear weapons program or against Iran's interference in Iraq and elsewhere.

Fourth, as demand for oil spreads around the world, and production of conventional oil is replaced by exploration and production in extreme environments, such as the Arctic and deep offshore waters, the distribution system—tankers, pipelines, oil storage facilities—becomes larger and more extended. This infrastructure is increasingly vulnerable to both natural and man-made disasters.

The 2010 BP oil spill in the Gulf of Mexico has led to greater environmental and economic damage than anyone contemplated. It certainly will lead to a major review of the risk of accidents by the U.S. government and oil companies that have oil and gas exploration and production (E&P) operations in the United States or elsewhere, especially in extreme environments. Reexamination of operating practices and regulations will likely take more than year, during which time new E&P deep water offshore operations will be curtailed, requiring adjustment to currently estimated quantities and cost of production. There is a danger that public attitudes and government policy will lead to an extended period of reduced investment and licensing, perhaps prompting similar caution in other countries, such as Brazil, which have deep water drilling operations.

Some observers will characterize the blowout accident as an exceptional case due to chance or to negligence; others will see the accident as evidence of general inattention to risk in the oil and gas supply chain. Few will recall that facilities in the Gulf of Mexico experienced significant damage from the 2005 Katrina hurricane, without appreciable interruption of production or damage to the environment.

The extended distribution infrastructure is subject to political interference, and it is highly vulnerable to attack by terrorists. Consumer and producer countries have a common interest in reducing the vulnerability of this infrastructure and in adopting plans for dealing with a disruption should one occur. Reducing the vulnerabilities of the infrastructure to human intervention also reduces environmental damage. For example, the oil leaks from Nigerian production are said to come almost entirely from acts of theft or sabotage.

There is sure to be a growing call for government and industry to pay greater attention to risks to the energy infrastructure, including accidents and safety, the ability to withstand natural disasters, and the possible threat of terrorist action, including cyber attack.

Fifth, international trade in natural gas is growing in importance. The expectation of more plentiful natural gas resources in North America, though, has reversed the assumption that natural gas imports to North America will soon become an essential and growing part of supply. But over time the international gas trade is sure to grow, accompanied by related import dependence security concerns.

Without change, the net effect of these trends in global oil and gas markets points to greater geopolitical tensions between three parties: the developed OECD importers, the rapidly growing emerging economies as they increase their already considerable demand in world oil markets, and the major resource holders. At best this situation will become a three-sided competition, as each party seeks to gain advantage. At worst, the situation will deteriorate, as economic competition in oil and gas markets turns into political competition for access to resources. There are many credible scenarios that could lead to political, and even military, conflict. For example, if political and economic conditions do not improve in some of the major resource holder countries, there is the possibility that these countries will experience internal upheaval resulting in new regimes with more extreme leadership. Nigeria is an example of a country where internal instability has interfered with orderly production operations. Some believe that uncertainty about access to oil could prompt action by outside forces to topple uncooperative governments of oil-producing states. China's inevitable appetite for oil and gas certainly adds strains within the region and to U.S.–China relations.

There are two additional points worth noting. The first is that geopolitics is not static; the political stability in the Persian Gulf region or other regions can change, perhaps substantially. Geopolitical change in turn will have an effect, either greater or lesser, on the significance of oil import dependence for U.S. foreign policy. Second, one can imagine that, over time, countries like Iran, Venezuela, and Iraq will become more democratic or at least more market oriented. It is tempting to assume that such a democratic trend, should it occur, would necessarily mean reduced concern with oil import dependence. Perhaps it would, but this is a proposition subject to question; a democratic regime can have sharply different interests from those of the United States, especially when it comes to energy policy. On the other hand, a radical takeover of an existing major resource producer does not necessarily mean inevitable disruption of supply. Radical governments also need export revenue to govern and finance their objectives.

"What If?" Scenarios

Speculating on the effect of hypothetical changes in the circumstances of oil import dependence sharpens an appreciation for the connection between import dependence and foreign policy. However, the limitation of this approach is that it is mere speculation with an outcome that cannot be validated. Here are three examples.

How would U.S. foreign policy differ if it were not dependent on imported oil? The question begs the answer that the United States would have an entirely different Middle East policy and a different military posture in the region as well. But the policy impact is quite uncertain because it depends upon other circumstances prevailing in the region that are not specified by the question. For example, would the decline in oil exports to the United

States necessarily impoverish Persian Gulf countries, such as Iran and Saudi Arabia? And how would these countries adjust to changed economic circumstances? Would other countries remain dependent on oil imports from the Persian Gulf? What would be the status of the Arab–Israeli conflict and of proliferation in the region? The point is, we know that oil dependence complicates the present political relationships and freedom of action of the United States. Absence of oil import dependence certainly would permit the United States to pursue other interests in the region and the world more freely, but the extent of the advantage cannot be precisely defined.

If the United States were not dependent on imported oil, would our force posture and defense expenditures be reduced and the need for military presence in the region eliminated? This question is related to the previous one. There is no question about the importance of Middle East oil to the world economy. In 1980 President Carter proclaimed the "Carter doctrine," which stated that the United States would defend its national interests in the Persian Gulf:

> Let our position be absolutely clear: An attempt by any outside force to gain control of the Persian Gulf region will be regarded as an assault on the vital interests of the United States of America, and such an assault will be repelled by any means necessary, including military force.[6]

The expression of this doctrine suggests that a principal purpose of U.S. military posture in the Persian Gulf is to ensure the security of supply, and that the fraction of the U.S. military budget associated with maintaining this military capability for intervening in the region should be "allocated" to the objective of security of oil supply.[7] The size of this military expenditure could then be compared to alternative policies for reducing oil import dependence, perhaps enhancing energy security at lower cost.

There are two serious flaws to this argument. The first is that there are many reasons for U.S. military capability and deployments in the Middle East, not just protecting the security of world oil supplies. Over the years these reasons have included preventing nuclear proliferation, dampening the potential for regional conflict with Israel, and deterring possible Soviet intervention across the Caucasus. So while it is clear that reduced dependence on Persian Gulf oil would not increase the need for the military capability to intervene in the region, it is by no means clear that reduced import dependence would reduce the military requirements in the region. Second, and more importantly, there is no theoretical or practical method to balance quantitatively the trade-off between measures that reduce imports with defense expenditures.

If oil prices declined sharply, what would be the effect on countries such as Iran, Venezuela, and Russia? These countries are dependent on oil revenues both for supporting their domestic economy and for financing their foreign activities. Clearly a significant decline in oil revenues would constrain the extent of their foreign activities and put pressure on their domestic economy. A cutback in the resources these countries can devote to foreign activities is surely welcome. But the consequence of domestic stress caused by fiscal pressure is less clear. In the extreme, the loss of revenue could destabilize the government and lead to regime change. The successor regime might be better or worse, from the U.S. viewpoint.

Is Energy Independence Possible?[8]

Given the recognition of the significant geopolitical costs of oil and gas import dependence, the initial impulse is to adopt an "energy independence" policy, seeking cessation of oil and gas imports in order to achieve the undeniable national security and alleged economic benefits of reliance on domestic resources only.

Most experts believe that the goal of energy independence is unattainable because of the high level of dependence on imported oil and gas in the U.S. economy, as indicated in Figure 3.1.

The level of imports has risen steadily over the past three decades; as of 2009, for example, crude oil and petroleum product imports account for about 60% of U.S. consumption. The most recent EIA *Annual Energy Outlook* (*AEO 2010*) rather optimistically projects, in the EIA's reference case, the net import share falling from 58% in 2007 to 44% in 2035.[9] The projected consumption level is essentially flat over the period because of the assumptions in the EIA NEMS projection system.[10] The projection includes a significant increase in domestic oil production, an increase in oil prices, the introduction of new energy efficient technology, and some change in consumer preference. For

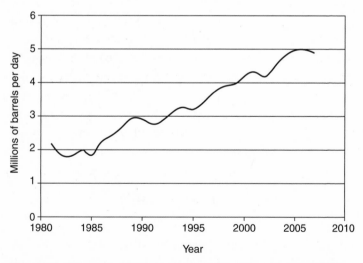

Figure 3.1. U.S. crude oil and product imports. *Source:* Data from EIA Web site, http://tonto.eia.doe.gov/dnav/pet/pet_sum_crdsnd_k_m.htm.

example, the 2010 *AEO* projects imported crude oil nominal prices to increase from a level of $90 per barrel in 2008 to $224 per barrel in 2035.

It is important to inquire about the possibilities for reducing this level of dependence and the economic cost of achieving import reductions. However, there is an accompanying question: How much of a reduction would be enough? As R. James Woolsey has stressed, a more nuanced meaning of "energy independence" is reducing import dependence so that oil is not a lever that exporting countries can use, explicitly or implicitly, to constrain U.S. foreign policy interests.[11] There is no objective calculation that can answer this question. Conceptually the criterion for an "acceptable" level of imports should be set by the magnitude of a supply disruption that the economy could absorb with low economic cost and disruption. In turn, resiliency to supply disruption depends upon the availability of alternative sources of supply, such as fuel switching and demand curtailment. My guess, and it is no more than that, is that that level might be 20%. Reducing petroleum imports from today's level of over 50% to 20% of liquid fuels in 2030 is an ambitious goal that would take significant money, time, and political courage.

Reduction in oil imports is achieved through either (1) reduction in total oil consumption or in proportion of imports as percentage of total oil use, as a result of adjustment to market conditions (for instance, fuel switching due to high oil prices), or (2) targeted import reduction measures, such as an import tax.[12] Because most oil use in the United States is for transportation, the possibilities are these:

1. Lower oil consumption due to higher oil prices (or poor economic conditions) or improved fuel efficiency from vehicle size, power plant improvements, or hybrid designs.[13]

2. Introduction of alternative liquid transportation fuels, such as biofuels or liquid fuels produced from natural gas, shale, or coal.

3. Substitution of natural gas as a transportation fuel.

4. Switch to hybrid electric or all-electric cars.

It is worthwhile to consider the effort needed to achieve a reduction in oil crude and product imports to a level of 20%, say by the year 2030. Table 3.1 includes a summary of the liquid fuel supply and disposition for the United States for the year 2030 compared to 2010, as presented in the reference case in the EIA's 2009 *Annual Energy Outlook*.

The EIA reference case projection is based on four important trends: (1) higher prices that will moderate demand growth, (2) introduction of flex-fuel vehicles that use alcohol/gasoline mixtures, including compressed natural gas, methanol, and ethanol fueled vehicles and compressed natural gas fueled vehicles, (3) increased market penetration of gasoline hybrid and diesel hybrid vehicles, and (4) introduction of plug-in hybrid and all-electric vehicles. The point is that the EIA reference case makes ambitious, if not heroic, assumptions about how much progress the country will make in reducing its consumption of petroleum as the economy grows and in introducing alternative transportation technologies.

Is there any prospect of doing better in reducing oil import dependence? And what might be the economic consequences of such improvement? Consider how the aspirational 20% import objective might be achieved by the year 2030. The third column in the table, marked "JMD aspirational scenario," indicates my speculation about how this level might be reached, using the EIA reference case as a starting point. This optimistic scenario is based on an expectation that the prospect for increased U.S.

Table 3.1. Oil and gas in the U.S. economy

U.S. liquid fuel balance MMb/day	EIA *Annual Energy Outlook* (2009) base case with stimulus		JMD aspirational scenario
	2010	2030	2030
Liquid fuel supply			
Domestic crude oil	5.52	7.14	7.50
Net crude imports	8.31	6.88	3.61
Net product imports	1.80	1.39	
NG plant liquids	1.87	1.87	2.00
Refinery gain	0.94	0.82	0.82
Other inputs	1.15	2.77	3.83
Ethanol	(0.85)	(1.72)	(1.50)
Biodiesel	(0.05)	(0.13)	(0.20)
Liquids from gas	(0.00)	(0.00)	(1.00)
Liquids from coal	(0.00)	(0.20)	(0.50)
Liquids from biomass	(0.00)	(0.33)	(0.63)
Total primary supply	19.59	20.87	17.76
Liquid fuel consumption			
LPG[1]	1.88	1.58	1.60
E-85[2]	0.00	1.20	1.50
Motor gasoline	9.43	8.24	6.74
Replace by NG vehicles			**(1.00)**
Replace by electric vehicles			**(0.50)**
Jet fuel	1.45	1.94	2.00
Distillate fuel oil	4.10	5.14	3.46
Replace by NG			**(1.68)**
of which (diesel)	(3.46)	(3.46)	(3.46)
Residual fuel oil	0.72	0.71	0.35
Other[3]	2.22	2.11	2.11
Total liquid consumption	19.80	20.90	17.76
Imported oil price $/b	78.00	125.00	
Natural gas price $/MCF[4]	6.05	8.40	
Imported oil as percent of liquid supply	52%	40%	20.30%

Source: EIA, *Annual Energy Outlook* (2009), updated for stimulus, table 11, "Liquids Fuel Supply & Disposition." Also appendix C1, "Price Case Comparisons."

1. Liquefied Petroleum Gases.
2. E-85 is 85% EtOH + 15% gasoline.
3. Other includes: aviation gasoline, waxes, lubricants, asphalt, heavy oils.
4. MCF—thousand cubic feet.

domestic natural gas supply is much more promising than is assumed in the EIA forecasts.

Experts generally agree with the sharply increasing estimates of natural gas reserves from "unconventional" resources, especially shale gas, in the United States and elsewhere in the world. In my judgment, natural gas substitution is the most likely, perhaps the only way short of draconian government regulation or a stagnant economy, to achieve the 20% objective. Accordingly, the JMD aspirational scenario assumes: on the demand side, significant substitution of natural gas for oil by (1) compressed natural gas vehicles, (2) substitution of diesel fuel in industry, and (3) replacement of residual fuel oil. In addition I am more optimistic about electric vehicles. On the supply side,[14] I have assumed: (1) significant growth in gas-to-liquids, (2) greater domestic production of petroleum, and (3) more production of coal-to-liquids and liquid fuels from cellulosic biomass. The basis for the aspirational scenario is the EIA 2030 forecast, but of course my assumptions are uncertain and open to objection. Perhaps the most ambitious assumption in the EIA 2030 scenario and the JMD aspirational scenario is the growth in domestic crude oil production. Crude oil production has been falling in the United States for many years due to the depletion of low-cost reserves, and the prospect for a reversal in this trend is dubious.

Caveats

The aspirational scenario achieves the 20% import objective by greatly expanding natural gas use. In that scenario, natural gas displaces 4.8 million barrels per day oil equivalent. This is roughly equivalent to 9.15 trillion cubic feet (TCF) per year. The EIA projects about 24 TCF per year natural gas domestic production in 2030 in the reference case, so it is clear that this scenario posits an expansion of domestic gas production over the EIA forecast of 38%. The assumption is that this expanded natural gas production comes entirely from North America, for if the supply were imported, say as LNG, the purpose of the reduction would be, to some extent, negated. On the other hand, because natural gas reserves are assumed to be plentiful in the scenario, the cost of the adjustment to the economy would be relatively low.

However, there is another point that deserves emphasis. U.S. energy security is tied to the energy security of its close allies and partners. Thus, if the United States were to reduce its petroleum imports to 20% of its liquid fuel use, but the European Union (for example, France and Germany) and developed Asia (for example, South Korea, Taiwan, and Japan) continued a high import level that influenced their conduct of foreign affairs, the benefits of U.S. energy independence would certainly be lessened. For the United States to realize the national security benefits of energy independence, it is necessary for both this country and its allies to reduce import dependence. The aspirational scenario presented above shows how difficult it is for the United States to achieve a 20% oil import objective; it is much harder for countries like Japan and Germany, which possess essentially no domestic oil reserves.

Conclusions

Energy security cannot be evaluated in exclusively economic terms—for example, by estimating the economic cost of a supply disruption of a particular magnitude and duration. Import dependence influences both directly and indirectly the foreign policy positions, leverage, and relationships of consuming countries and the influence, capacity, and options of producing countries. Energy, and especially oil, is an instrument of geopolitical power.

My conclusion is that "energy independence," even in the limited sense of an "acceptable" 20% oil import target, is an unrealistic objective. Although all would welcome it, I conclude that in practice, at least for the next several decades, it will not be possible for the United States, or its allies, to reduce import dependence to acceptably low "aspirational" levels. But import levels do matter. From the national security perspective, less imports is better, and there are reasons other than import reduction to adopt government policies that encourage reduction in petroleum use, for example, beginning the long-term transition from dependence on fossil fuels, foreign or domestic.

Accordingly, the United States must manage energy security in a realistic manner that ensures that both the international repercussions of domestic energy policy actions and the domestic energy consequences of foreign policy actions are central to all policy deliberations. The appropriate advice to our national leaders is to prepare for managing difficult crises, and perhaps even conflict, in the years ahead. It is most likely that the United States and other countries will remain dependent on oil and gas imports for many decades. This means that every importing country will need to balance the security disadvantages of imports with the economic advantage of lower-cost oil. Although it is sensible to adopt policies that will reduce this dependence over time, it is futile to seek to eliminate entirely or artificially constrain the use of these resources, on which the economies of both producing and consuming countries depend. At the end of this book, in Chapter 6, I offer some recommendations as to how this mutual interdependence can be managed, as well as policies that should be considered for reducing dependence over time.

4

BIOMASS, SOLAR, AND NUCLEAR ENERGY, WITH AN ASIDE ON NATURAL GAS

One of the major energy challenges facing the country is the transition from an economy based on fossil fuels to an economy based on nuclear and renewable energy sources. The transition is necessary in order to avoid the environmental consequences of burning fossil fuels, to mitigate progressively higher costs anticipated for oil and gas as lower-cost resources are exhausted, and to address security concerns that accompany dependence on oil and gas imports.

The ideal energy technology is cheap, plentiful, and environmentally friendly and job creating. But good intentions are not sufficient for a new technology to succeed. It is unlikely that any new technology can simultaneously satisfy all these criteria. As my colleague Lester Thurow observed: "It is only when we demand a solution with no cost that there are no solutions."

Several energy sources have the potential to replace a significant portion of the massive amount of fossil fuels on which the United States and other countries depend. Given human imagination and our capacity for innovation, it is not surprising that many potential sources have been suggested. In this chapter I shall discuss three of these alternative sources: solar, biomass, and nuclear. I do not address geothermal, ocean waves or tides, wind, and many others that are under investigation. Moreover, I address only some aspects of solar, biomass, and nuclear energy—in particular, solar electric and solar-to-fuels (not solar thermal applications), biofuels from carbohydrates and cellulosic feedstock (not algae, for example), nuclear fission (but not breeders or fusion). Thus, my discussion is far from inclusive. My intention is to illustrate the complexities of integrating the political, economic, and technical considerations required for successful innovation.

Gasohol and the Prospects for Liquid Fuels from Biomass Feedstock

In 1979, when I was director of Energy Research in the DOE, I learned that President Carter was planning to support a legislative proposal, sponsored by senators from agricultural states, for a 4¢ per gallon tax credit for gasohol, a mixture of 90% motor gasoline and 10% domestically produced ethanol—an alcohol made by fermentation of corn or sugar. The justification for the tax credit was that gasohol domestically produced to replace motor gasoline would, at the margin, displace imported oil.

The DOE staff pointed out to me that because U.S. agriculture was highly industrialized (which was not true everywhere—Brazil was an example), a significant amount of oil and natural gas was employed in the cultivation of the corn, in the subsequent fermentation of the corn into ethanol, and in separation of ethanol by distillation from the fermentation mash. As a result, the net amount of imported oil displaced by gasohol would be significantly less than the gross amount of gasoline displaced by the ethanol. In sum, it seemed likely that this politically popular proposal was an extremely expensive way to reduce the amount of foreign imported oil for the U.S. economy. The U.S. taxpayer would be subsidizing ethanol production and paying a higher cost than necessary to displace imported oil.

I proceeded, as might any technical person, inexperienced in the influence of special interests in Washington, to form a group of experts under the external DOE Energy Research Advisory Board (ERAB) to look into the matter and report the facts to me and to Charles Duncan, the secretary of energy. The committee, chaired by Professor David Pimentel of Cornell University, was composed of university and industry experts on fermentation, agricultural practice, and chemical processing.[1]

The findings of the ERAB committee are summarized in Figure 4.1 (taken from the 1980 ERAB report). Approximately 45,000 BTUs of oil and natural gas (primarily for fertilizer and fuel for agricultural machinery) are required for cultivation of the corn necessary to produce one gallon of ethanol.[2] An additional 56,000–70,000 BTUs, depending on the efficiency of the fermentation and distillation process, are needed to convert the corn into ethanol. Distillers dry grain (DDG) is an important side product; seven pounds of DDG are produced from the 20 pounds of corn required to produce one gallon of ethanol. Because the DDG is as good a cattle feed as corn, the DDG should be taken as a by-product credit, which reduces the cultivation

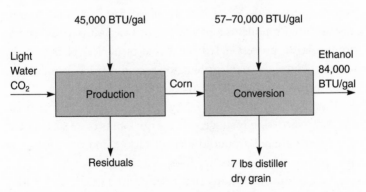

Figure 4.1. Findings of the ERAB committee, 1980.

energy amount from 45,000 to 30,000 BTUs per gallon of ethanol produced. Ethanol's energy content is 84,000 BTUs per gallon compared to gasoline's energy content of 125,000 BTU per gallon, so a gallon of ethanol has an energy equivalence of two-thirds of a gallon of motor gasoline.

The result of the analysis is that one gallon of ethanol requires 100,000 (30,000, with DDG credit, + 70,000) BTUs of premium fuel energy (oil and natural gas), or 16,000 more BTUs than the 84,000 BTUs the ethanol will displace.[3] In the simplest terms, premium energy inputs are greater than the premium energy value of the produced ethanol, even when by-product credits are taken into account.

Gasohol advocates were outraged by this ERAB report. It led to three congressmen, including Tom Daschle of South Dakota (later the Senate majority leader), to request an inquiry by the congressional watchdog, the Government Accounting Office (GAO), about the conduct of the DOE gasohol study.[4] The Office of Technology Assessment (OTA) was asked to mount a competing analysis.[5] Controversy about the appropriate numbers persists to this day.

The ERAB analysis assumed that diesel oil or natural gas was used to produce the heat required for fermentation and distillation. Proponents argued that the heat source could come from solar or process heat from a coal or nuclear plant, which would reduce the requirement for premium fuel. Perhaps so, but this was not the practice then, and it is not now.

In 2002 the Department of Agriculture (not a likely gasohol skeptic) issued a study suggesting energy inputs of 23,861 BTUs per gallon ethanol produced for cultivation, and 51,779 BTUs per gallon ethanol produced for conversion.[6] Including by-product credit, this means that corn-derived ethanol displaces 16,314 BTUs per gallon of ethanol. Because the 4¢ per gallon tax credit applies to the 90/10 gasohol mixture, the subsidy is

40¢ per gallon or $16 per barrel ethanol. If the ethanol displaces about 16,313 BTUs of motor gasoline per gallon, or about 650,000 BTUs per barrel (and we ignore any change in volume on mixing), then the subsidy equates to the government paying $122 dollars per barrel of imported oil displaced, a very hefty price to pay compared to other measures that could reduce reliance on imported oil, such as spending money to improve the efficiency of energy use. Many states offer additional state gasoline tax credit for gasohol, which increases the incentive substantially.

Gasohol prospers in the market because of the subsidy and the presence of a renewable fuel standard enacted in the 2005 Energy Policy Act that mandates gasohol as part of the motor gasoline pool of the United States.[7] Ethanol in motor gasoline is projected to increase from 9.84 billion gallons in 2008 to 34 billion gallons in 2035, an annual growth rate of over 4% during a period when the motor gasoline pool is projected to remain constant.[8]

My effort in 1979 to slow subsidies to gasohol backfired. What are the lessons? First, no issue can successfully be resolved if it is assumed to be completely a technical matter. The politics of other interests cannot be ignored. Second, it is not possible to fight something with nothing. In this case, the alternative is cellulosic biomass, as this feedstock does not come from the energy intensive agricultural sector that produces corn, sugar, and other foodstuffs.[9]

Cellulosic Biomass

Sugars are linked in a 'α' or 'β' pattern. The 'α' linkage gives sugars that are easily metabolized by fermentation bacteria and digested by living organisms. These sugars are a source of food. The 'β' linkage sugars give rise to celluloses that are difficult to metabolize and hence are unsuitable for food. Examples of cellulosic biomass are cellulose, hemicellulose, corn stovers, lignin, agricultural waste, perennial grasses such as miscanthus, and switch grass and other plants that grow wild without energy input for cultivation. For this category of biomass, the "premium" energy balance is not an issue.

However, making ethanol from 'β' linkage is more difficult, because this linkage packs the sugar molecules more closely, yielding a denser feedstock. Harsh energy intensive chemical pretreatment, as indicated in Figure 4.2, is required to prepare the dissolution and fermentation. Nevertheless, the "premium" energy balance is significantly more favorable. More importantly, the cost of producing ethanol from cellulosic feedstock appears to be significantly lower than the cost of producing ethanol from corn or sugar. The cellulosic feedstock cost is mainly for collection, while the corn feedstock cost depends on its food value, which more than offsets the higher capital cost of the chemical pretreatment process step.

Since 2005 the DOE has embarked on an impressive research program for cellulosic biofuels.[10] The effort includes genetic

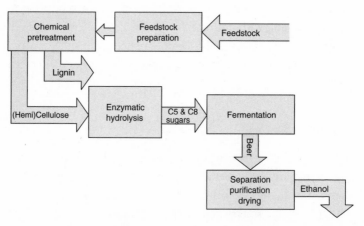

Figure 4.2. Making ethanol.

engineering of the initial plant material to produce higher yield, alternative plant structures, and production of chemicals within the plant leaf. Work is also under way on new chemical pretreatment techniques and bioengineered enzymatic organisms with better fermentation performance.

In December 2009 the Obama administration announced that $600 million would be allocated to projects to demonstrate the commercial potential of advanced biofuels (most of these funds come from the stimulus package).[11] In addition, private venture capital is backing dozens of start-ups that are developing new approaches to producing biofuels, from thermochemical processes to genetic engineering of plants and bioengineering more efficient fermentation.

Today energy legislation recognizes the advantage of cellulosic feedstock over corn and offers preferential assistance terms to cellulosic feedstock biomass projects. So it seems that broader energy reasoning is now displacing narrow agricultural interests (although an unjustifiable protectionist import duty on imported ethanol remains).

I believe biofuels have considerable promise as a liquid transportation fuel. Although biofuels derived from corn and sugar do not make sense in the United States, this is not necessarily the case for Brazil or other countries with a more favorable climate and a less energy intensive agricultural sector.

For cellulosic biofuels, the question in the United States and elsewhere is, of course, the cost and timing of significant cellulosic biofuels production. Absent subsidies, I estimate that cellulosic liquid biofuels might be produced for about $50 per barrel, compared to synthetic liquids from coal or shale with CCS in the range of $70 to $100 per barrel.

There are various estimates of ultimate global penetration of biofuels (up to tens of millions of barrels of oil equivalent per year annual production), but such production levels raise other questions. The first concerns the availability of water. Biofuel production requires significant amounts of water, and in many places, such as China and Africa, water supplies are scarce. Second, while cellulosic biomass is not directly a food crop, land use for extensive cellulosic biomass production might indirectly impact food production. Third, the assumption that biomass production creates no net CO_2 increase, because harvested crops are replaced by new planting, is oversimplified. For example, a shift in the mix of crops and crop rotation will change soil conditions and, in particular, the soil moisture content. This, in turn, influences the flux of CO_2 that is exchanged between land and atmosphere and the productivity of the land. The net effect of these changes must be included in the CO_2 offsets that are attributed to biofuel use.

The ultimate technical biofuel goal is to find a synthetic way to reproduce the photosynthetic process of plants: the transformation of sunlight plus CO_2 and water to useful chemical products. Plants do this reliably, but slowly and rather inefficiently; their main trick is to gather sunlight and use the photon energy to transfer electrons to oxygen, thus reducing oxygen to a chemically potent agent. A forward-looking research program should include artificial photosynthesis and photochemical conversion of water to hydrogen and oxygen as a high-risk, but potentially high-payoff, effort.[12]

Lessons Learned

The gasohol story is a vivid and not pretty example of what happens when technology mixes with politics and how special interests that have political influence can determine unsound energy policy. A credible biofuels path has emerged, and both industry and government are sensibly pursuing it. But much time has been lost, and billions of taxpayer's dollars have gone into activities that are of little consequence for the country's energy future.

Photovoltaics: Sunlight to Electricity

The photovoltaics story differs from that of gasohol; it is not special interests but rather widespread hope that dominates the tale—the search for an economical technology that, once built, will produce electricity without any fuel other than sunlight and with no undesirable environmental effects.

Device Physics

The physical principle of photovoltaics (for economy, I refer to photovoltaics as PV in this section) is that photons impinging on a semiconductor device will separate positive charges and electrons that produce direct-current electricity, if the photon energy exceeds the energy band gap of the semiconductor material. Sunlight is composed of photons with a distribution of energy; the sun's "spectrum" is this distribution of photon energy. The greater the fraction of the sun's spectrum that is converted to electrical current by the device (see Figure 4.3), the higher the efficiency of the device per unit area.

Many clever ways have been invented to increase PV efficiency, including new materials, multilayered materials designed to absorb photons in different parts of the solar spectrum, anti-reflection coatings on the front of the device, and reflective coating at the back of the device. A variety of materials are in use: crystalline silicon, amorphous silicon, cadmium telluride, gallium arsenide, and copper indium gallium selenide (CIGS). There is considerable research on the use of thin films, flexible structures, and nanotechnology. In sum, PV is an exciting research and exploratory development field. There has been steady

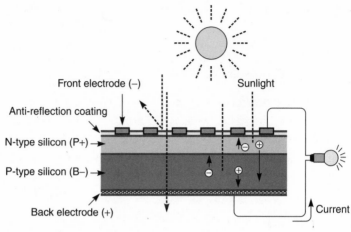

Figure 4.3. Sunlight to electricity.

progress in improving device efficiency, as Figure 4.4, on the best research cell efficiencies, indicates.

PV as a System

The purpose of PV is to produce electricity, either in distributed generation from rooftop installations serving residential customers, or from a large PV installation (hundreds of megawatts) that is connected to the electricity transmission grid. In order to assess properly the performance and cost of a PV source, it is necessary to consider the complexities of the system application; these complexities unfortunately temper the prospects for PV. The performance of a laboratory device is necessary, but not sufficient, to ensure successful practical technical and economic performance.

The first reality is that sunlight impinging directly on the active surface of a solar panel depends on latitude, season, and time of day. Second, as illustrated in Figure 4.5, the PV system consists

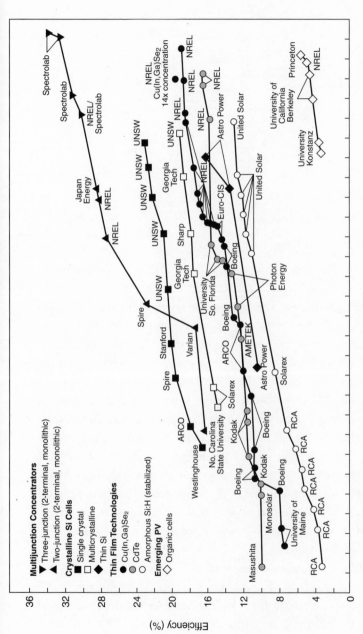

Figure 4.4. Research cell efficiencies. *Source:* Chart from the DOE's NREL National Center for Photovoltaics, available at http://www.nrel.gov/pv/thin_film/docs/kaz_best_research_cells.ppt.

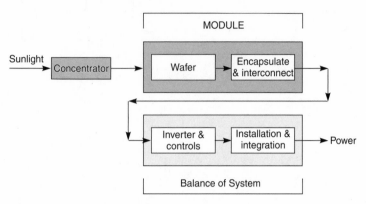

Figure 4.5. Photovoltaic system.

of a module and a balance of system. The modules contain an interconnected array of the solar cells that converts the sunlight to DC electricity. The balance of system includes conversion and control for the DC-to-AC power and the integration and installation into the distribution system. The PV system may also include fixed or moving solar collectors to concentrate the sunlight impinging on the solar cells.

Thus, PV includes more than solar cells, and the cost of the electricity depends on more than cell efficiency. To be sure, achieving higher cell efficiency is desirable because it reduces the area and hence the cost of producing electricity. But some of the costs—such as balance-of-system costs, which can amount to 50% of the capital cost of the PV system—do not depend on area. The point is that science will get you only so far. Engineering and technology are also needed for successful application of photovoltaics.

PV system capital cost is expressed as $ per peak watt (W_p).[13]

Because PV systems have very low operating costs, the "levelized cost of electricity" for PV can be estimated from the capital cost, the annual cost of this capital (say 10%), and the number of kilowatt-hours that the system rated for W_p (peak watt) will produce in a given site on an average day. The levelized cost of electricity refers to the electricity cost per kWe-h of electricity produced by a system throughout its life needed to cover all expenses in the construction and operation of system; the levelized cost allocates the costs of building and operating the system to the power produced by the system during its operating life.[14]

In midlatitudes, about 1 kilowatt of direct sunlight falls on each square meter of surface. The efficiency of the PV module at converting sunlight to DC electricity determines the area of the module that must be designed and manufactured for a rated value, say 1 kilowatt peak electricity. On a typical day this device might produce 4 kilowatt electric hours (kWe-h) when sited in a midlatitude location. If the cost of an installed grid-connected PV system is $1 per peak watt, then the levelized cost would be about 7¢ per kWe-h, a cost comparable to the range of traditional base load power at the point of generation (without a carbon emission charge).

However, the $1 per W_p is an ambitious cost target.[15] A good deal of renewable energy policy is concerned with how to get the cost substantially lower—I address this matter below—but the elements are clear: increase efficiency to reduce area-dependent array cost, reduce manufacturing cost, and reduce balance-of-system and installation cost.

Intermittency

It is widely and incorrectly believed that if the levelized cost of solar would drop to the point of other grid-connected electricity-generating technologies, then solar would rapidly penetrate the electricity-generating market. There are several reasons levelized cost is not the only factor that determines the relative economic value of solar.

First, investments in new technology are made on the basis of "net present value" (NPV)—the discounted present value of the difference between revenue and cost over future time periods. If a solar PV system generates its power during time periods when the users demand load and hence the prevailing electricity demand is high, and if local regulations permit the PV system access to this market to supply its power, the NPV may be attractive—the higher revenue offsets relatively high generating costs.

Second, because solar is carbon-free and does not depend on fossil fuels, it has a social benefit that is not necessarily recognized by the market. This benefit justifies public assistance for solar, nuclear, or other renewable technologies, such as wind and geothermal, which may lift the NPV of a photovoltaic project significantly.

The third reason, intermittency in solar power generation, is less widely appreciated. Intermittency of power output has significant adverse cost implications for solar power (and other intermittent sources, such as wind) at high penetration levels in an electrical grid. Solar power generation varies with season, time of day, and local weather conditions. Some of these variations can be predicted, but some, such as those due to local weather, are more uncertain. Unlike base load conventional power generation, such as a coal plant that can increase its dispatch of power in response to an increase in demand, PV produces electricity strictly in proportion to sunlight and cannot dispatch its power in response to change in load. PV power generation is, thus, intermittent.

At low penetration, the grid can accommodate the intermittency. At high penetration, say greater than 10%, the overall system must make changes to accommodate intermittent solar power. Possible measures to compensate for intermittency are electricity storage, additional backup power generation, and expanded transmission capacity to shift available generation to meet load requirements in response to loss of supply of intermittent power.

Intermittency also limits the capability of distributed PV systems to meet demand. Here the choices are to arrange for backup power by grid connection, add local storage (for instance, batteries), or establish a hybrid system (such as a local natural gas electricity generator). For photovoltaic systems that are deployed in either distributed to grid-connected configurations, each of these measures represents an additional cost of operating the electricity system beyond the levelized cost of the PV system, and this additional cost of intermittency should be included in estimates of solar cost at large penetration.

Moreover, intermittency has implications for the design of a "smart grid" that is intended to encourage more efficient use of electricity. Optimizing grid design requires consideration of several variables: the characteristics of different generation technologies, storage options to lessen the effects of intermittency, transmission capacity, and demand-side management practices, including time-of-day pricing. The importance of grid modernization is recognized today, but most proposals address change to the existing electricity system, one of these variables at a time, within its regulatory framework. This process will not achieve system optimization, because interactions between the variables are not considered. The practical consequence of this suboptimal approach is slower progress toward achieving economic efficiency for the electricity system.

Government Policy to Accelerate Solar Deployment

It is clear that the public wants the government to encourage the use of solar energy. But that imperative does not determine the nature of the policies or the amount of support the government should extend for this purpose. I address the justification for government policies to assist new energy technologies in Chapter 5. Here, I focus on government assistance to PV, an important new solar energy technology.

There is strong theoretical justification for government support for research and development, the early stage of innovation, and for the past three decades the DOE has had a program for that purpose for PV. The PV R&D program includes materials research (ranging from crystalline silicon to thin polymer films to quantum dots), multijunction high-efficiency cell design, concentrators, manufacturing and measurement and characterization.[16] I regard this R&D effort as generally successful at creating new knowledge important for exploiting the PV technology options. The effort would have been more successful if the PV R&D program had not experienced ups and downs in funding, reflecting different attitudes toward energy under different administrations.

The more difficult question is the nature and extent of initiatives the government should take to encourage the demonstration and the deployment of photovoltaics at the stage when the technology is not economical compared to other available electricity-generating alternatives. The purpose of demonstration projects is to establish the technical performance, cost, and environmental characteristics of a technology. If the technology demonstration projects are relatively inexpensive, in range of a few million or even tens of millions of dollars, the private sector has the capacity to take on the project risk in the hope of capturing early mover benefit. If the demonstration projects are of large size, say in the range of $1 billion or more, and if there is significant regulatory uncertainty (for example, with regard to first-of-a-kind nuclear power, carbon sequestration, and smart-grid projects), the private sector cannot reasonably be expected to take on the risk and some government assistance is necessary.

PV projects are not of this scale, but nevertheless the U.S. DOE and states have supported many PV demonstration projects, reflecting the public preference for solar technology.

Another point of view is that government should take steps to accelerate deployment because the market does not internalize the external cost of the alternative electricity-generating technologies, such as GHG emissions. A variety of mechanisms are available for this purpose: production incentives or Renewable Portfolio Standards. Germany, for example, offers very generous feed-in-tariffs, as high as 10¢ per kWe-h for PV electricity production, that are intended to, over time, reduce the unit cost of PV systems due to learning and economies of scale. Several states have adopted Renewable Portfolio Standards that require utilities to generate a certain fraction of electricity by PV or other renewable technologies. If a serious GHG emission charge were in place, it is unclear that these additional subsidies would be necessary or justified.

Nuclear Energy

Concerns about climate change and other environmental problems of coal and the future availability and price of natural gas have led both developed and developing countries to renewed interest in nuclear power. In 2003 I co-chaired, with Ernie Moniz, an interdisciplinary MIT faculty study titled *The Future of Nuclear Power.*[17] In 2009 we issued an *Update to the MIT 2003 Future of Nuclear Power Study.*[18] The main objective of these studies was to provide an objective assessment of the prospects for nuclear power both in the United States and around the world.

Nuclear power has had a long and difficult history. Initial high expectations for nuclear power have not been met; nuclear power is not "too cheap to meter."[19]

Reactor Safety

Perhaps the dominant reason that much of the public and many political leaders have become disillusioned with nuclear power is primarily because of safety concerns caused by the accidents at Three Mile Island, Harrisburg, Pennsylvania, in 1979, and at Chernobyl, Ukraine, in 1986, and to the lack of progress in radioactive waste management.

On March 11, 2011, a severe earthquake and tsunami hit the Tokyo Electric Power Company (TEPCO) Fukushima Dalichi power station. The event, in strength well beyond the design-basis threat, created serious damage to the six reactors and associated spent fuel storage pools at the site, due to loss of cooling water caused by the disruption of all external and backup power. As I write this in April 2011, considerable time and effort is being devoted to restore the plant to a safe condition. Less damage was sustained by the nearby Fukushima Daini power station, which hosts four reactors.

It is most unlikely that nuclear power deployment around the world will be unaffected by this accident. The calls for a moratorium on new construction and license extension are likely to overwhelm the protestations that nuclear power is safe. The public will expect and deserves an objective assessment of the implications of the accident for the present and future use of nuclear power. Serious accidents such as the Fukushima or the Macondo oil spill in the Gulf of Mexico that resulted from an explosion on the Deepwater Horizon drill rig operated by BP in April 2010, are reminders that major energy activities inevitably bear a safety risk whether from natural disasters, human or equipment error, or sabotage (including cyber terrorism).

Risk is the product of likelihood and consequence of a severe accident. The public and regulatory agencies are willing to accept safety risk, but nuclear accidents are particularly worrisome because of the possibility of many fatalities from radiation release—a risk with a chance of 1/10,000 of 10,000 fatalities is less acceptable than a risk with a certain outcome of one fatality, even though both have the same expected outcome. Accordingly, licensing of nuclear facilities focuses on analysis of a plant's ability to withstand postulated events, such as an earthquake of a certain strength that is judged to have some small probability of being exceeded, say one in a million during the next hundred years. If the low likelihood event occurs, it is wrong to conclude that either the procedure is bad or that the limit has been set too low. However it is reasonable and appropriate to inquire if the process of setting the limit and estimating the likelihood was properly followed, and if new measures should be taken to reduce the consequences of an accident at existing and future facilities should they occur.

This process of review will take place in the United States, Japan, and other countries. The review will involve a number of different topics, for example, provisions for backup power, consideration of siting of multiple reactors at one site, continued reliance on storage of spent fuel in pools at reactor sites rather than in dry storage at a central location, and emergency preparedness and planning for accident response by the utility and public authorities. Such a review will take considerable time because of the many factors involved and the need to analyze the balance between the benefits of additional safety measures with their costs. Moreover, there will need to be opportunity for public comment as well as extensive regulatory review. Here again is an example of the need to integrate technical, economic, and environmental views in order to resolve an energy issue.

The Fukushima accident has significantly reduced the prospects for the growth of nuclear power during the coming decade, thus reducing, at least for the near term, an important low-carbon electricity generating option.[20]

Proliferation

An additional concern is that the spread of nuclear fuel cycle technologies, enrichment, and reprocessing could increase the risk of the spread of nuclear weapons capability. Indeed, nuclear energy is the most vivid example of what can go wrong in technology innovation when understandable enthusiasm for a genuine new technical opportunity gets wildly ahead of the economics, regulatory requirements, and public understanding.

Today nuclear power faces four challenges: (1) building power plants at lower capital cost so that the cost of electricity production is comparable to the cost for natural gas and coal, (2) making progress on radioactive waste management, (3) convincing the public that operation of nuclear power plants is sufficiently safe, and (4) minimizing the risk of proliferation from the spread of nuclear power to other countries.

The essential hurdle is to lower the cost of building new nuclear power plants. All indications (including foreign construction experience) are that the estimated capital cost of new reactors is too high, in excess of $4,000 per kWe of capacity[21] and increasing faster than the competing base load technologies, such as coal and natural-gas-fired electricity generation. This corresponds to a generating cost of 8.4¢ per kWe-h compared to 6.2¢ per kWe-h for coal and 6.5¢ per kWe-h for natural gas (with $7 per thousand cubic feet natural gas fuel cost), exclusive of any carbon charge. The *MIT Future of Nuclear Power Study* advocated federal support for building a limited number of new nuclear plants in order to establish for investors that nuclear plants could be successfully licensed and built at acceptable cost.

The MIT study favored assistance in the form of production tax credits or payments, as an incentive for project completion, but instead Congress extended loan guarantees, presumably because of the lower budget impact.[22] But loan guarantees, in contrast to production incentives, insure projects from failure, rather than incentivizing success. In February 10, 2010, the DOE announced $8.33 billion in loan guarantees to a group of Georgia utilities for construction of two 1,110 MWe Westinghouse AP1000 reactors at an existing nuclear station site in Burke, Georgia.[23] The size of this loan guarantee implies a very high project capital cost, which is troublesome because the future of nuclear power in the United States depends crucially on demonstrating that the technology can produce electricity at low cost.

The outlook for waste management is also not promising. President Obama decided to abandon the costly Yucca Mountain geological nuclear waste disposal project, thus eliminating any near-term prospect for resolving the fate of spent nuclear fuel accumulating at commercial reactor sites. In January 2010, DOE Secretary Steven Chu announced formation of a commission co-chaired by former Indiana congressman Lee Hamilton and former national security advisor Brent Scowcroft to provide recommendations on managing used fuel and nuclear waste. As this is only the last of a series of commissions, by itself it is not an indication that a credible framework for nuclear waste management will be established soon.

The nuclear safety outlook was more positive until the March 2011 Japanese nuclear accident. All technical experts agree that the design and operating practices of the new generation of nuclear reactors is an order of magnitude better, measured in terms of the estimated mean years of reactor operation to failure relative to the current fleet, thus supporting a larger fleet size without incurring the risk of more accidents.[24] In the United States and elsewhere, operating nuclear reactors have in general achieved progressively higher utilization rates, from an average of about 70% in the 1990s to over 90% in 2002.[25] This utilization improvement increases the effective generation of the fixed nuclear reactor fleet; and, as Admiral Rickover was fond of saying, "an operating reactor is a safe reactor."

The proliferation issue is complicated and vexing, and has a long history. Interest in nuclear power is particularly strong in countries around the world that anticipate rapid growth in electricity demand—for example, Indonesia, Turkey, Egypt, Jordan, Abu Dhabi, Taiwan, South Korea, Chile, Argentina, and Brazil. The question is how best to extend the benefits of nuclear power while avoiding the risk that nuclear material produced in the fuel cycle might be diverted to weapons application. The two critical activities are enrichment and reprocessing of spent fuel.

Nuclear power requires fuel with a higher concentration of the uranium isotope U^{235} that is split (fissioned) by neutrons to produce energy. Enrichment refers to the process that increases the abundance of the U^{235} isotope from its natural amount, 0.7%, to the 3% to 4% required for nuclear fuel. There is a variety of enrichment technologies. Today gaseous diffusion and centrifugation are in use; tomorrow, perhaps separation by selective laser irradiation. However, enrichment plants can be relatively easily reconfigured to produce higher enrichment—to greater than 20%, the level that, by convention, is considered "highly enriched" and hence potentially suitable for a nuclear explosive device.

Spent nuclear fuel from a reactor contains small amounts of plutonium (directly usable on nuclear devices), formed by neutron absorption of the nonradioactive, abundant U^{238} isotope, followed by emission of an electron. Plutonium is a different atomic species from uranium and thus can be separated by chemical means. The separation process, originally designed during World War II, is called PUREX and relies on solvent extraction of plutonium. Solvent extraction is a process in which a species, in this case plutonium, is selectively dissolved in one liquid from another immiscible liquid. The process is dangerous because of the exposure to the highly radioactive waste stream, but it is simple and well known.

Enrichment and reprocessing technologies are proliferation risks, and the risk justifies export controls and other measures to restrict the spread of these dangerous technologies while not restricting access to commercial nuclear power.

Figure 4.6 indicates the places of concern in the nuclear fuel cycle.

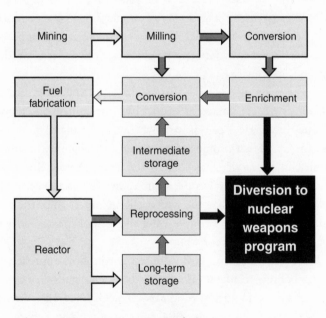

Figure 4.6. Places of concern in the nuclear fuel cycle. *Source:* "The Nuclear Fuel Cycle," Center for Nonproliferation Studies, Monterey, Calif., http://cns.miis.edu/research/wmdme/flow/iran/index.htm#1.

Countries seek a nuclear weapons capability because they perceive that possession of nuclear weapons meets a security need. So it is unrealistic to believe that stopping access to nuclear power means that countries will be prevented from acquiring nuclear weapons. Indeed, all countries that are believed to possess nuclear weapons capability, with the exception of India, obtained the strategic nuclear material (highly enriched uranium or plutonium) from dedicated, clandestine facilities.[26] However, possessing commercial fuel-cycle facilities puts a country closer to a bomb capability and in a position to elect to develop a weapon capability in stages. Iran's example shows how a country can build facilities for enrichment for commercial nuclear power that also have the capacity to produce nuclear material needed for a nuclear device.

The developing countries that are expected to be new users of nuclear power are interested in pursuing enrichment and reprocessing for a variety of reasons. Without enrichment, countries are concerned (with some reason) about the reliability of supply of enrichment services provided by the United States, Russia, or Europe. Reprocessing has the allure of being an advanced technology, to better utilize the fission energy potential of the uranium resource, and, of course, offer the unspoken possibility of obtaining plutonium for a bomb.

In recent years the G-8 adopted a common approach to strengthening proliferation safeguards in anticipation of a possible expansion in nuclear power use. The G-8 has taken a series of measures beginning in 2004 and announced its support of new mechanisms for nuclear supplier states to supply fuel cycle services to states that want to use nuclear power.[27]

The essence of the policy is that countries that do not possess uranium enrichment and plutonium reprocessing facilities would agree not to obtain any such facilities and related technologies and materials. In exchange, they would receive guaranteed cradle-to-grave fuel services under an agreement that was financially attractive and signed by all those countries in a position to provide them. The International Atomic Energy Agency (IAEA) would sign also and would apply safeguards to any such fuel cycle activities covered by the agreement, in addition to its traditional safeguard activities with regard to the reactors in the recipient states. The idea is to make costly indigenous fuel cycle facilities less attractive than reliable fuel cycle services from a few nuclear supplier states.[28]

To be effective, the arrangement would need to include a guarantee to the recipient country, at least for enrichment services. It is not likely that a recipient country would make an investment in a reactor without an international guarantee that a quarrelsome U.S. Congress could not abruptly terminate commercial contracts for enriched fuel, for example. The guarantee would be strengthened by an international enrichment "bank" or "reserve," operated perhaps by the IAEA. In a remarkable display of public support for such a security initiative, former U.S. senator Sam Nunn, the co-chair of the U.S.-based Nuclear Threat Initiative (NTI), announced that American entrepreneur Warren Buffett had pledged $50 million toward a total of $150 million for a low-enriched uranium stockpile owned and operated by the IAEA.[29]

This is a new approach that amounts to an important revision of the terms of President Dwight D. Eisenhower's 1953 "atoms for peace" deal between nuclear have and have-not states. The implicit deal in "atoms for peace" was that nuclear weapon states would provide access to technology and nuclear material to non-weapon states in exchange for the recipient non-weapon state agreeing to forgo nuclear weapons. The proposed conditions for today's deal are more stringent: recipient states get access to technology and nuclear power reactors but not to the more sensitive parts of the fuel cycle. These fuel cycle services—enrichment and reprocessing—would come from a restricted number of nuclear supplier states. The universe of nuclear supplier states remains to be defined; presumably, supplier states initially would consist of the existing nuclear weapon states plus some others, such as Germany and Japan.

Of course, a good policy idea may be a long way from a functioning policy. As yet there is no concrete example of such an arrangement. At one point the Iranian Bushehr-1 reactor, constructed by the Russians, seemed to be a good model. The Russians would "lease" the fuel—that is, they would provide enriched fuel and take back the depleted fuel for reprocessing or disposal, or both. However, the unwillingness of the Iranians to suspend their enrichment activities at Natanz has halted any progress on international support for Iran's nuclear program.[30]

Brazil's plan for construction of a $210 million enrichment facility at Resende (a project run by the Brazilian navy) presented a second opportunity to achieve this new fuel cycle arrangement.[31] The United States chose to acquiesce to the Brazilian enrichment program and to object to the Iranian enrichment program on the grounds that the latter was dangerous and the former was not. I do not believe a proliferation policy will be workable in the long run unless the rules have international scope and can be consistently and objectively applied. The Brazilian decision to proceed with its domestic enrichment plant is a setback to the proposed G-8 policy to internationalize fuel cycle services.

A third event was the December 2006 signing of the U.S.–India Peaceful Atomic Energy Cooperation Act.[32] Although the strengthening of the political relationship between the United States and India has much to recommend it, this action is not positive on nonproliferation grounds because it serves to legitimize a country that is not a Nonproliferation Treaty signatory; India is instead an "undeclared" nuclear weapons state that does not permit IAEA inspections of its nuclear facilities, which include PUREX reprocessing plants of significant capacity.

Practical realization of the new policy calls urgently for one example that works. An important, infrequently acknowledged barrier is that few countries are willing to accept the return of spent fuel. Russia will take back fuel of Russian origin; the situation with regard to the European Union and France, in particular, is less clear. There is little indication, despite its strong commitment to nonproliferation, that the United States would be willing to accept returned spent fuel; perhaps this attitude will change when an effective nuclear waste disposal system is in place.

The G-8 nonproliferation initiative that offers to provide assurances of fresh fuel and spent fuel management to states that agree not to pursue enrichment and reprocessing programs deserves strong support.

The Reprocessing Question

The United States presently uses an "open" fuel cycle, or "once-through" fuel cycle, where the spent fuel from reactors is not recycled but is discarded in a geologic repository with its long-lived (greater than 10,000-year lifetime) actinide isotopes. A "closed" cycle separates the plutonium and reuses it as reactor fuel where it produces energy when it fissions. The alleged advantages of the closed cycle are these: (1) It makes the waste management task easier, because the absence of long-lived isotopes means the nuclear waste's radioactivity decays sooner (after several hundred years, rather than several tens of thousands of years), (2) the uranium resource base is extended manyfold because of the breeding and reuse of plutonium in power reactors, and (3) the new system can be made proliferation resistant.

The United States had a long debate in the 1970s about the relative merits of a closed versus open fuel cycle. President Ford canceled plans for reprocessing of commercial spent fuel in 1975. President Carter placed the country on the open fuel cycle path, canceled projects related to the closed fuel cycle—for example, the Clinch River breeder reactor in Tennessee—and argued in international forums, such as the International Fuel Cycle Evaluation study, that the closed fuel cycle, as then configured, presented unacceptable risks in international commerce because it made bomb-usable separated plutonium widely available for commercial nuclear power. The U.S. view influenced the attitudes of nuclear supplier countries, such as France, Germany, and the United Kingdom, about proliferation risks of fuel cycle exports, but did not convince these countries to abandon the closed cycle indigenously. For example, France is the world leader in the development and operation of commercial nuclear fuel reprocessing at La Hague. Japan and the United Kingdom have had mixed results in their reprocessing efforts.

The debate about the merits of reprocessing continues. Some believe the most important advantage is for waste disposal, because of the long-term environmental benefits of removing long-lived actinides from the waste; this presumes that the separated actinides are burned in reactors. Of course, the long-term benefit must be balanced against the near-term risks of operating complex reprocessing and fuel fabrication plants. All agree that the closed cycle will be more expensive (although not a large percentage of the total cost of electricity) than the open cycle for many decades, until there is sufficient deployment of conventional nuclear power plants to drive up the cost of natural uranium ore to the point where reprocessing of spent fuel for use in advanced fuel cycle is economical.

In my opinion the strongest objection to the United States reprocessing commercial spent fuel is the consequence for the

U.S. nonproliferation strategy. At the same time that the G-8 is attempting to convince other countries not to deploy indigenous reprocessing technology, nuclear supplier states are seen as pursuing the technology for themselves. Other countries, such as Iran, Brazil, Turkey, South Korea, and Taiwan, might well wonder whether they are being asked to give up an important technology.

The alleged improved "proliferation resistance" of the advanced reprocessing technologies that could be developed (at substantial cost) is a fantasy. The only countries that have the scale and technical sophistication to adopt the elaborate new advanced fuel cycle configurations that link reactors and advanced reprocessing plants are nuclear supplier states, such as France, Russia, and the United States, that have passed the point of being proliferation risks. New user nations will not have a nuclear power industry of sufficient size to justify the large and expensive advanced fuel cycle processes developed by supplier states; these countries will select PUREX (as India is doing) because it is known and simple.[33] In any case, if a country decided to divert material, relatively modest additional processing would be needed to recover pure plutonium.

The George W. Bush administration supported work on advanced fuel cycles that avoid separation of pure plutonium as having a greater proliferation resistance. The vision was to develop, through a Global Nuclear Energy Partnership (GNEP), an advanced closed fuel cycle that is based on a proliferation-resistant fuel cycle that makes it as difficult as possible to misuse or divert nuclear materials to weapons. The GNEP advanced fuel cycle concept avoided separation of pure plutonium by keeping it with uranium and co-located separation plants with reactors that would burn the fissile plutonium/uranium mixed oxide fuel. My view is that such an effort will in fact derail the prospects for an orderly expansion of nuclear power throughout the world at a time when there are few alternatives to further emissions of CO_2 from electricity-generating technologies.

In time, if use of nuclear power significantly expands around the world, it may be justifiable to adopt a closed fuel cycle. But at present there is no compelling reason to do so. Some developed nuclear supplier countries, such as France, Japan, and Russia, will choose to continue to pursue a closed fuel cycle domestically; others, such as the United States, Germany, and Canada, will wisely choose to avoid reprocessing and the closed cycle for the foreseeable future.

There is a political dynamic in the United States that makes the closed fuel cycle attractive. The U.S. Congress is weary of cost overruns and delays in constructing the waste repository at Yucca Mountain, Nevada. Congress has the false hope that a closed cycle will be easier to accomplish and politically more acceptable. But opposition will mount quickly. Proliferation risk is a powerful public issue; developing a closed fuel cycle will cost billions of incremental research and development dollars that many will judge could be better spent on other energy programs, especially on developing technologies that encourage greater energy efficiency. Finally, siting reprocessing plants and the ancillary fuel fabrication facilities will prove to be no easier than siting waste disposal sites, once the risks associated with operation of the fuel cycle facilities are understood by the public.

Conclusions

The discussion of these three technologies—biofuels, solar, and nuclear—illustrates the complexity of deploying new sources of energy supply. Technical, political, and economic elements are part of the story, and progress depends on wise and consistent government policy. The biofuels example illustrates that avoiding the political reality is a mistake; the solar example illustrates that avoiding the reality of the linkage between technology readiness and economics is a mistake; the nuclear example illustrates the importance of taking into account the international aspects of energy technology. In short, oversimplifying the complexity of these technology deployment challenges does not make progress surer, it makes progress less likely. How best to manage technical innovation in such an environment is an important question, to which I turn in the next chapter.

5

MANAGING ENERGY
TECHNOLOGY INNOVATION

Federal government support of innovation—of both the creation and the demonstration of technology—encourages private investors to adopt new technology. In this chapter I review the success of the Department of Energy (DOE) in advancing technology for the energy sector. I argue that the DOE and it predecessor agencies have had better success in the early stages of innovation (sponsoring R&D to create new technology options) than in the later stage of innovation (demonstrating technologies with the objective of encouraging adoption by the private sector). The DOE does not have the expertise, policy instruments, or contracting flexibility to manage successfully technology demonstration. I suggest that it is necessary to establish a new mechanism for this purpose.

Virtually every energy study recommends that the federal government support technology research, development, and demonstration (RD&D) programs that require large and sustained budgetary support—paid, of course, by the taxpayer. In previous chapters I have discussed government programs for carbon capture and sequestration and for renewable technologies, such as photovoltaics and wind, biofuels, and nuclear power.

Every advocate for each of these technologies is genuinely convinced of the merit of his or her approach for achieving desirable technical change and the justification for government assistance. However, particular interest groups or constituencies, whether farmers, university researchers, or private firms, often lack candor about the benefits they receive from the RD&D projects they champion.

We should strive to understand which government efforts are likely to be effective and which are likely to be ineffective in encouraging adoption of new energy technologies. The government must decide which of the many candidate RD&D programs to pursue, how large a program to mount, and how best to manage the effort.

Innovation is the process by which technical change is accomplished. The innovation process consists of two steps: the first step is technology creation—the discovery of a new science or technology. The government, private industry, and foundations sponsor discovery activities. Industry, universities, and both federal and not-for-profit laboratories and hospitals perform the R&D.

The second step is deployment of the new science and technology into an enterprise or the society. This is, by far, the more difficult step in achieving technical change, because it usually involves: (1) making an uncertain investment decision; (2) managing change in a production process, along with its workforce; and (3) tailoring a new service or product to customer need.

Nations and firms that do innovation well have an advantage over their competition and enjoy greater economic growth. Innovation has as its objective both improving performance at some given cost and or achieving some level of performance at lower cost. Finally, in the case of fielding a new technology to accommodate new environmental regulations, the objective of the innovation is to maintain output while meeting more stringent standards, and at roughly the same cost as before the regulation.

The Government's Role in Encouraging Innovation

The government has three functions in the innovation process: the first function is to set the rules for the innovation activity. Setting the rules enables innovation and determines whether the innovation process will perform well or not. The rules include defining intellectual property rights; setting and publishing standards for materials, products, and safety; establishing tax treatment for the cost of R&D activities undertaken by private firms; controlling export of technology; creating mechanisms for industry/university/government partnerships; and providing access to venture capital.

The importance of the rule-setting function is frequently overlooked. However, countries that set the innovation rules "right" do a lot better than those who do not. My impression is that the United States and Japan generally have set the rules right, although the two countries have quite different approaches. Europe has been less successful at nurturing start-up companies and encouraging collaboration between industry and universities.

The second government function is to support technology creation. The justification for this role is well founded, especially for the early stage of the discovery process. Uncertainty as to the eventual realization of long-term benefits from fundamental research means that private firms are not assured of capturing these benefits and so will invest less than what is optimal for the society. When early-stage "pre-competitive" technology creation is supported by the government, on the other hand, the results are made available to all.

The U.S. government has proven most successful in the technology creation phase. The federal government spent $114.6 billion for all R&D in 2008. Of this amount, $55.1 billion was devoted to basic and applied research. The DOE's share of these expenditures was $8.2 billion for all R&D and $6.2 billion for basic and applied research in 2009.[1] The data do not include the very significant addition to federally sponsored R&D included in the stimulus package; the American Recovery and Reinvestment Act added about $18 billion to federal R&D obligations in 2009.

Federal support to basic and applied research and for the creation of research facilities has a long history in this country. No other nation has remotely as successful an R&D enterprise, and the rest of the world emulates our practices. The hallmarks of the U.S. approach are project selection according to merit, and, in general, flexibility in accommodating education as an important by-product of funded research activity. Universities, government laboratories, and industry perform the basic and applied research; university-based research is particularly strong in the United States compared to other countries. The DOE government laboratories focus on basic energy science, energy technology, high-energy and nuclear physics, and nuclear weapons research, development, test, and evaluation. The DOE operates many large research facilities for a wide range of scientific users groups. The government manager who successfully fosters technology creation is knowledgeable about advances in the field and attentive to outside expert opinion; direct support of R&D projects is the manager's major tool.

The third function of government is to support technology demonstration. Here the government has had a good deal more difficulty in successfully influencing the process of transfer, adoption, and deployment of new technology. The closer the government-sponsored activity comes to demonstrating a potentially useful commercial product, the more difficult it is to justify spending taxpayer money instead of relying on private market decisions. Moreover, how should benefits be shared when the government supports a private firm in demonstrating the practical application of a technical advance? For example, should the company that receives government funding for a demonstration project own the intellectual property? Or should the intellectual property remain with the government to sell or distribute to future users?

Another approach the government can take to encouraging technology demonstration and deployment is to rely on regulatory mandates. The basic concept, usually applied in an environmental control context, is for the government to specify a schedule by which firms must comply with specified performance (such as end-of-pipe emissions) standards. The mandate is seen to "pull" new technology into the marketplace. There are many successful examples, including vehicle fuel economy and air and water quality standards. The approach, however, is not based on demonstrating the uncertain technical performance, cost, or environmental effects of a new technology. Rather, regulatory mandates serve to establish certainty with regard to future industry, making compliance not a matter of choice. Indeed, in order for the regulatory mandate to be successful, it must be based on a known and reliable technology, known as "best and available technology." The mandate successfully pulls technology into the marketplace by requiring industry to adopt it. To be sure, once the regulatory mandate establishes a new economic reality, there is an incentive to develop (and demonstrate) improved technology, but this is a secondary consequence. Regulatory mandates are a mechanism for setting a policy constraint and should be compared to a tax, cap-and-trade, or other market based mechanism.

Commercial and Government Markets

The government encourages deployment of new technology in two different markets. In the first market, the government is the sole customer for the new technology. The traditional examples are the nation's defense, intelligence, and space programs. For this market, the problem of technology deployment is simple, because the government runs the activity. The desired technical change does not have to meet a commercial market test, but rather needs to meet performance goals established by the government. Examples are: NASA's Mars landing program or the DOD's effort to modernize military functions. In this market, the major uncertainty facing the government manager is whether a technology project will meet set performance, schedule, and cost objectives. Of course, the cultural hurdle of convincing existing institutions to accept change is present, but the uncertainties in price and competition associated with a large private market are not.

History shows that the United States has been quite successful in utilizing technology for government activities and achieving the second step of the innovative process, for example, in exploiting technology for the military. To be sure, the process may be spectacularly expensive. An internal resource allocation process applies some discipline to the entire activity and gets the job done.

It is important to appreciate that, in practice, much government funded technology creation to support public activities has an enormous range of unplanned benefits to the commercial economy. For example, DOD supported technical advances on network communications, computer systems, and solid state electronic devices, motivated by military applications, are largely responsible for today's modern information technology society. The United States enjoys a great advantage from the flexibility that this "dual-use" pattern provides—an advantage that other nations, for example, the Soviet Union, were unable to exploit.

For defense and aerospace R&D the government is the sole sponsor and the principal user.[2] The motivation for innovation for these sectors is "technology push," higher performance with less emphasis placed on cost. New technology does not need to compete in a market where the adoption depends upon price compared to available alternatives. The defense/aerospace industry is composed of a relatively small group of national firms that compete for government contracts according to a very formal process. The paradigm for this type of technology development is "linear," with passage through sequential stages: research, exploratory development, advanced development, engineering, manufacturing and production, and field support, for example, in the case of high performance combat aircraft. The R&D process and the defense budget is managed according to this linear paradigm.

Innovation in the government market is a very poor model for how innovation takes place in the private sector, so we should expect a very different model for government supported innovation in sectors such as energy or health that operate in a commercial market.

The government objective in supporting energy technology demonstration is to foster investment by the private sector in select technologies. In the energy sector innovation a new technology competes in the market on the basis of price and performance compared to existing or potential alternatives. Here the motivation for innovation is "application pull," and cost is paramount. The innovation process paradigm is "parallel" rather than linear, as consideration of technical performance, engineering, design for manufacturing and cost must proceed simultaneously, right from the beginning of a project. System engineering—a discipline rarely taught in universities—plays an even more prominent role than in the defense/aerospace sector. So when a disruptive technical idea is invented, a complex process is needed to transform the idea into a practical and economic system. University based research typically is not experienced in this transformation process.

The private sector will adopt new technology only when it believes the innovation will be profitable, under anticipated market conditions. Thus, if the government hopes to encourage adoption of new technology, the government program must take into account the uncertainties associated with a private market, for example, market prices and regulatory conditions, e.g., emission controls. Prices send signals for both the supply and demand of the products and services. Accordingly, price uncertainty will influence the relative economic attractiveness of candidate new technologies, and thus price is another uncertainty that must be considered in the R&D process.

Because of the size of the energy enterprise, vast amounts of capital and long times are required for the deployment of new technology, e.g., a "smart" transmission grid, an electric car, a coal plant with capture and sequestration of the produced CO_2. In the United Sates and the developed world, private capital finances energy projects, and it is not surprising that private investors, before committing large amounts of capital, want to see practical demonstration on a commercial basis of the technical performance, cost, and environmental characteristics of the new technology. This represents a difficult extension of the government role beyond paying for R&D performed by universities, industrial firms, and not-for-profit laboratories to create new technology options, to demonstrating first-of-kind performance.

A wider range of tools is useful for technology demonstration than the conventional direct government payment for R&D to support the earlier discovery phase. These indirect mechanisms include loan guarantees, tax credits, and guaranteed purchase, production payments. Indirect mechanisms are especially suitable for technology demonstration because they interfere minimally with commercial practice in project design and management, thus providing more credible evidence to private investors on the commercial feasibility of the technology. In contrast regulatory mandates, such as renewable portfolio standards (RPS) for electric utilities or biofuel renewable fuel standards, create a market but without requiring commercial viability. An advantage of both indirect mechanisms and regulatory mandates to support technology demonstration is that they avoid the cumbersome and expensive procurement regulations required for all government procurement activities that have no parallel in the commercial world.

The Government Record in Demonstration of New Energy Technology

The government has a mixed record of achieving desired technical change in the private sector. The National Institutes of Health has been remarkably successfully in fostering advances in the biomedical sciences and transferring this knowledge and associated technology to both big pharmaceutical companies and small biotechnology companies born from NIH-funded research at universities, medical schools, and hospitals. Over the years, the Department of Agriculture's extension service has successfully transferred technology and know-how to the American farmer, enabling a vast increase in agricultural productivity.

The record of the Department of Energy and its predecessor agencies, however, is decidedly more mixed. Here are some examples.

- Nuclear power has received special attention from DOE, its predecessor agencies, the Energy Research and Development Administration, ERDA, and the Atomic Energy Commission, AEC, because the technology originated exclusively from the government weapons program. While there were some notable technical successes, most knowledgeable observers would consider that the effort failed, especially with regard to nuclear waste disposal and high capital cost discussed in Chapter 4.

- Beginning in the 1980s, the DOE launched a program focused on clean coal technology that operated by competitive selection of strictly cost-shared industry projects and was granted multiyear funding by Congress. While there were some successes, for example, development of sulfur dioxide scrubbers and selective catalytic reduction, the results of this effort were mixed, because many of the projects had more to do with subsidizing a particular company's development than demonstrating new technology.

- Government-funded demonstration projects (sometimes conducted with industry partners) frequently have been over budget and conveyed little useful information to the private sector. Examples include the Clinch River Breeder Reactor, the Barstow Solar Power Tower, and several synthetic fuel plants have a particularly poor record.

- On several occasions, the DOE has undertaken smaller scale demonstrations, e.g., photovoltaic, wind, and fuel cell projects. However, these efforts are more a response to congressional interest than a serious attempt at technology transfer.

- The DOE has experimented with supporting industry consortia on the reasonable ground that industry managed efforts have a greater chance to cause technical change in the private sector. Examples include support for the Gas Research Institute (GRI, now abandoned), the Advanced Battery Consortium (ABC), the Partnership for a New Generation of Vehicles (PNGV), and encouraging (but not directly funding) the Electric Power Research Institute (EPRI). Each of these efforts has made some contribution, but none has been sufficiently successful to suggest adopting consortia as a general model.

- Periodically, federal purchase programs are proposed, for example, for natural gas or electric vehicles, as an effective way to demonstrate new technology. More problematic are proposals for buy-down campaigns (for example, purchasing photovoltaic modules), as an effective way to drive unit costs of new technology down to economic levels.

- Federal and state subsidies, usually in the form of tax credits for favored technologies, such as wind and bio-fuels, are offered as an effective way to promote energy technology. The rationale for this approach is that using public money to provide information to the private sector about the economic, technical, and environmental performance of new energy technology will lead to widespread adoption by industry.

On one occasion, the government mounted a much larger scale attempt to introduce technology that would change the course of energy development in this country. The significance of this case is that it was the only effort that approaches the scale of government action that many believe is necessary today.

Lessons from the Synthetic Fuels Program

I ask you to recall the infamous Synthetic Fuels Program, launched in 1980 and ignominiously abandoned in 1986. The lessons of this experience go beyond the usual criticism by censorious economists of government involvement in technology commercialization.

The Energy Security Act of 1980 established the U.S. Synthetic Fuels Corporation[3] (SFC) at the height of the oil crisis for the purpose of establishing a domestic industry to produce synthetic gas and liquids from tar sands, shale, and coal as an alternative to oil imports. At the time of the SFC debate, oil prices were about $40/barrel and seemed to be headed for $80–$100/b. With little relevant experience, engineering estimates were that synthetic fuels would cost about $60/b. Accordingly, there was significant political pressure to demonstrate a domestic synfuels production capability that would act as a "backstop" to the seemingly endless upward movement of imported oil prices. Congress, industry, and a surprising number of informed energy and international security experts argued that the proper way to demonstrate this "backstop" price was to establish a production target: 500,000 barrels per day for the first phase.

The initial "first of a kind" plants were expected to cost more, justifying a larger subsidy to begin the "learning" process that many believed would result in lower costs. As late as 1982, in the Reagan administration, the DOE estimated that synfuels production in 2000 could be between 474,000 and 3.2 million barrels per day oil equivalent.[4]

The subsequent sad story is well known. In fact, the price of oil did not go to $100 per barrel but rather tumbled to less than $20 per barrel (Figure 5.1). The SFC struggled on, managing a handful of projects, until it was terminated in 1986.[5] Most of the projects selected by the SFC were brought in on schedule but at a cost vastly above the prevailing market price.

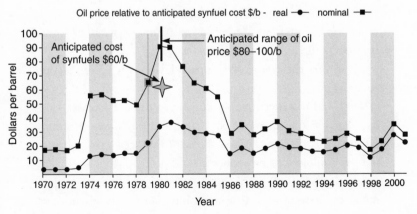

Figure 5.1. Cost per barrel of oil, 1970–2000.

The most charitable, but wrong, characterization of the principal lesson of the SFC is that the mistake was to misestimate future oil prices. There are many aspects of the SFC that can be criticized, but to condemn the basic rationale because the price of oil fell, is like faulting someone for buying an insurance policy, paying the premium, and then living. It is not a mistake per se to buy insurance or a hedge that later proves to be out of the money.

The primary lesson of the SFC story is that the government should be very cautious in establishing large programs based on the assumption that current estimates of market price will come to pass. The potentially expensive word "demonstration" should be carefully defined to avoid adopting either production targets or fanciful buy-down or learning ideas independent of real market experience and unexpected political, regulatory, and technical events. The SFC experience would have been more successful or, at least, less expensive, if "demonstration" had meant providing information to the private sector on the technical, environmental, and cost of a synfuels technology, rather than attempting to achieve production targets independent of the prevailing market price for conventional oil and gas. The SFC experience warns against adopting formulaic policies such as "renewable portfolio standards" and arbitrary emission reduction targets that are based on assumptions about future market conditions.

However, the SFC offers other lessons that are relevant today:

First, indirect incentives—production payments or tax credits, loans or loan guarantees, guaranteed purchase—are more effective for "demonstrating" to the private sector that a particular technology can be economic and profitably deployed. The alternative of direct DOE involvement in the design and payment for the cost of a demonstration plant[6] is simply not credible to the private sector.

Second, the strength of federal support for R&D lies in the earlier stages of innovation, especially in creating the basis for new technology. Government procurement rules are not germane, and the expertise of government R&D managers is not relevant to the decision-making required for investment under uncertainty that is at the heart of the commercialization phase of a new technology.

Third, large energy outlay programs attract more than normal congressional interest. Understandably, members like to have the projects in their districts and seek to influence the DOE decision-making process. The long and checkered history of the FutureGen integrated coal-fired generating plant with carbon capture and sequestration project in Illinois is a vivid recent example. A quasi-public corporation, such as the SFC, insulates the program to some considerable degree from congressional pressures and the annual budget cycle.

The Way Forward

Should the federal government undertake a program to encourage the commercialization of energy technology by industry? The preceding discussion explains why it is not surprising that while there is broad agreement about the reasons for government concern with energy innovation, there has been much less agreement about the federal role in the later stages of demonstration and commercialization of energy technologies. Many observers believe that these later stage innovation activities require the federal government to make a judgment about future winners and losers in the private marketplace. There is considerable skepticism that the DOE can effectively make such judgments, because the government bureaucracy lacks the necessary skills, and the agency is subject to short-run congressional interests. On balance, I favor government involvement in technology demonstration, but it is important to understand the limitations and risks in this activity, and that technology demonstration is only one aspect of what the government should do to encourage energy technology innovation.

Limitations

There are several factors that explain why these demonstrations have been less successful than hoped:

Lack of adequate resources for demonstration purposes. Except for the brief period of the SFC in the early 1980s and the current period of budgetary plenty accompanying the stimulus package, Congress has been restrained in voting funds for demonstration projects. The exception is the politically powerful clean coal constituency, but even cost sharing conditions are motivated more by a desire to reduce outlays than by any logic on how costs should split between the government and private entity. Demonstration projects rarely are required or allocated sufficient resources to undertake analysis of operating experience that can provide valuable information to investors, which is, after all, the purpose of technology demonstration. Finally, the large price tag of a demonstration project, often over several billion dollars, means it is always worthwhile to inquire if information from a smaller process development unit project would not provide much of the key information that is sought.

Currently, the DOE is funding a number of demonstration projects through a loan guarantee program established by title XVII the 2005 Policy Act. Significant additional funds, $6 billion, were made available under the American Recovery and Reinvestment Act in 2008 to extend the loan guarantees, and awards that have been made to nuclear power, $8.3 billion, and central solar power plants, $1.37 billion. Other assistance programs funded under the American Recovery and Reinvestment Act address coal plants with CCS, alternative transportation vehicles, battery manufacturing plants, and creation of "smart" electricity transmission grids. While the funds were freely flowing in 2010, it is not unlikely that that as economic recovery happens, concern with the growing fiscal deficit will sharply curtail these demonstration activities. I cannot avoid the impression that projects are being funded without sufficient attention to the core issues of specifying the criteria that should be applied in selecting projects and specifying the data that should be collected and disseminated from these projects to encourage future deployment.

The DOE's capacity to manage technology demonstration projects. We should be realistic about the capacity of the DOE system to manage technical innovation. The Department's strength in technology management is with R&D—the discovery phase of the innovation process. Technical program managers can rely on the considerable expertise that resides in the Department's laboratory system. Appropriated funds directly support the cost of the R&D, so there is reasonable control over the work effort, whether performed by government laboratories, universities, or industry.

On the other hand, how well can the DOE meet the criterion for a technology commercialization success? For a first-of-a-kind demonstration, the criterion is whether information obtained about technical performance and cost influences private sector investment decisions. As I have mentioned, the DOE has no expertise at making investment decisions under uncertainty that is key to private sector innovation. It is unreasonable to believe that the DOE, or indeed, any government agency, can develop this expertise in-house or (as has been attempted from time to time) contract for it. But, there are other hurdles as well. The federal and DOE procurement rules and management practices make it difficult to structure a demonstration project that is credible to the private sector. The DOE is accustomed to financing projects by paying directly all or a portion of project cost, and it does not have experience or authority in the use of indirect incentives, such as guaranteed purchase or favorable financing, which might place a demonstration project, for example, a photovoltaic production plant, on a commercial footing.

Most importantly, the success of any commercialization project requires a stable source of funding on a set project schedule. Frequent changes in direction mandated by a new administration or a congressional committee are unhelpful. Finally, the DOE and its oversight committees in Congress are continually lobbied by special interests—coal, carbon, California—who argue for projects that benefit their industry, community, or public interest constituency. Under these circumstances, it is almost impossible to adopt and sustain an objective and analytically based energy technology commercialization strategy.

Advantages of the U.S. System

I want to note that the United States has several potential advantages over other countries in pursuing energy innovation: (1) availability of sophisticated modeling and simulation capability that greatly assists design, (2) systems engineering capability in industry that is vital for economical and reliable operation of energy systems, (3) a project management capability unmatched by any other country, (4) an exceptionally strong basic and applied research base that, compared to other countries, is well coupled to commercial industry, especially energy.

These advantages are recognized both here and abroad. This means that U.S. industry is seeking to develop advanced energy technologies in all areas from renewable energy to carbon capture and sequestration with the intention of exporting the technologies to other countries. Developing countries, especially China, see the combination of U.S. diplomatic initiatives advocating an agenda of change for energy and the environment, especially emission reductions, and the prospects of U.S. energy technology exports as a U.S. effort to gain commercial advantage. Accordingly, access to energy technology has become an active topic in international negotiations, once again illustrating the domestic/international linkage of energy issues.

The strength of the United States in science and engineering underpinning energy technology means that U.S. participation is valued in international collaborations in fusion energy, nuclear energy, and basic science activities that involve large research instruments, such as high energy and nuclear physics.

Alternative Mechanisms for Managing Energy Technology Demonstration[7]

I conclude that a successful government program to demonstrate new energy technologies requires the establishment of a new mechanism significantly different from the current DOE program approach. To be successful the new mechanism must be able to:

- provide indirect incentives in order to make the demonstration as credible as possible to private investors;
- rely on commercial practices free from the government procurement rules that govern funding of R&D projects;
- have access to adequate, multiyear funding that permits efficient and orderly execution of the demonstration projects.

How might such a new mechanism for selection and management of projects that receive government assistance be organized?

It is conceivable that a separate unit within the DOE similar to the current DOE loan guarantee office[8] will work, if given these authorities, but I doubt it. The responsibility to demonstrate energy technology might be spread among several departments—the DOE, the Department of Commerce, and the Department of Agriculture—but this would spread responsibility, authority, and resources too thinly.

Some years ago, economics professor Paul Romer of Stanford University suggested an interesting approach of relying on self-organized industry investment boards that would operate as a bank to finance projects of collective interest.[9] There have several recent legislative proposals for "green banks" that would provide financing to energy concerns with renewable or energy efficient products.[10]

There are two other approaches worth mentioning. The first is to create a new Department of Industry and Technology that would combine the R&D functions of the DOE and the National Science Foundation with the technology demonstration and commercialization responsibilities of the DOE, the Department of Commerce, and other agencies. The scope of such a department would be larger than just energy and would run the danger of unwisely blurring the distinction between R&D and technology demonstration.

The second approach is the creation in the DOE in 2007 of the Advanced Research Projects Agency–Energy (ARPA-E),[11] the analog of the Defense Advanced Research Projects Agency (DARPA), which is charged with funding potentially "transformational" energy technologies. ARPA-E is strongly supported by Steven Chu, the current secretary of the DOE. Initially the ARPA-E concept did not strike me as especially compelling. The DOE innovation process, as described previously, is different from the DOD process in that DOE must struggle with commercialization and private markets. Moreover, ARPA-E is intended to address disruptive technology creation rather than the more difficult step of technology commercialization.

However, in its short history ARPA-E has proven extremely successful and made a number of interesting innovations in R&D management. ARPA-E solicited proposals from all performers, but with an emphasis on start-up companies spun out of universities labs. The administration of the solicitation, evaluation, and selection process was streamlined compared to the process for the normal DOE science offices, and an astonishing number of firms responded to the opportunity. I was impressed by the many interesting new approaches to key problems, such as energy storage and solar, selected in the first round of ARPA-E awards. The ARPA-E experience reminds us that new approaches to R&D management can yield unexpected benefits.

An Energy Technology Corporation

I prefer an approach that creates a separate, quasi-public corporation—the Energy Technology Corporation (ETC)—that is based on the best features of the SFC.[12] The ETC would select and manage technology demonstration projects without favoring particular fuels or supply. Congress would establish the ETC with a board of directors responsible for its governance. The board of directors would require Senate confirmation, and the board would have the authority to name the officers of the ETC and set their compensation.

The objective of the ETC would be to select and support demonstration of large-scale key energy technologies—projects that demonstrate the technical performance, cost, and environmental effects of new energy technology. The ETC could provide assistance through direct grants to industrial companies and industrial consortia, with or without cost sharing. In addition, the ETC would have the authority to make loans to and direct investments in commercialization ventures under terms (including accepting ownership interest) that the board deemed reasonable. Thus the ETC would have considerably greater flexibility than government agencies to encourage commercialization. The advantage of this approach is that the new corporation could operate in a manner closely analogous to the private sector, unfettered by federal government rules concerning people or procurement.

The ETC would be composed of independent individuals with experience and knowledge about future market needs, industry capability, and best use of indirect financial incentives—loans, loan guarantees, production tax credits, and guaranteed purchase—in order to run a project on as commercial a basis as possible. The ETC would not be subject to federal procurement rules, and if financed with a single appropriation, would be somewhat insulated from congressional and special interest pressure. The key difference between the SFC and the ETC is that the ETC would undertake demonstration projects to produce information and not produce predetermined output quantities. The information would guide the future investment decisions of private sector entities (and the banks that finance their activities); therefore the charter of the ETC would need to be carefully drawn up to reflect these criteria.

The disadvantages of the corporation alternative are that it requires creating an entirely new entity that would not build upon the capability and interests of existing agencies, it would be difficult to sell to Congress, and its performance is highly uncertain. This is the high-risk, high-payoff alternative path. In sum, however, the ETC would have considerably greater flexibility than the DOE in executing technology demonstration projects.

It does not make much sense to establish such a mechanism unless the scale of the effort is substantial, such as capital in the range of tens of billions of dollars to be committed over a period of several years. This amount would permit the ETC to provide sufficient financial incentives (but not to pay the entire cost) for a range of technology demonstration projects (for example, those identified in Title XVII of the 2005 Energy Policy Act establishing loan guarantee authority for the DOE) for (1) carbon capture and sequestration projects, (2) photovoltaic module fabrication, (3) new nuclear plants, (4) electric grid modernization, (5) central solar power projects, (6) fuel cells for residential, industrial, or transportation use, (7) cellulosic biofuels plants, and (8) hybrid/electric vehicle manufacture.

I believe the ETC is a better mechanism than the industry energy investment boards or energy banks mechanisms because ETC focuses on demonstrating technical performance, cost, and environmental effects of selected projects, while industry boards suggests financial assistance for deployment with a less clear statement of the criteria of selecting projects. Furthermore, the ETC explicitly envisions a large onetime commitment of federal resources to cost-share demonstration projects. The investment board/bank approach seems exclusively to rely on private sector funding, consistent with a deployment focus. Of course, it is possible to modify each proposal so that it resembles the other, so an intermediate solution might be preferable; for example, the industry investment board suggests an even greater private sector voice in selecting projects. The main point is to provide preferential financing assistance with some public resources to demonstrate new energy technology.

The ETC would not sponsor R&D or process development unit-scale engineering projects—these activities would remain the responsibility of the DOE. Thus the ETC would not support carbon capture and sequestration science but would finance a carbon sequestration demonstration project.

The ETC would function as follows: First, it would specify areas of interest and the kinds of assistance mechanisms the

Corporation would consider. Second, the CTC would be required to solicit proposals, evaluate submissions, and select winners. Third, the ETC would need to implement its investments and follow the progress of its projects. Fourth, the ETC would be required to submit to the president and Congress an annual evaluation of the Corporation's progress.

This brief description avoids many thorny design issues, such as who would own the intellectual property created by the project, and whether non-U.S. firms should be eligible to participate.

Some observers will note that this design for the ETC is similar to the approach to the synthetic fuels program described above. But the SFC was given production goals, not a technology demonstration mandate. The "failure" of the synthetic fuels program was not the result of the mechanism. In fact, the SFC operated fairly well, and the projects the SFC supported were completed on time and on budget. The synthetic fuels program "failed" because crude oil prices were sharply lower than expected and below the level that made production economic.

Conclusion

The DOE's role in managing energy innovation goes well beyond sponsoring basic research and early technology development, although this is where the government involvement has greatest theoretical justification and the DOE's record of performance is strongest.

The difficult challenge is to understand how the DOE best encourages commercialization of new energy technologies. The private sector is slow to deploy new technologies because of the risks involved. The risk goes beyond the uncertainty about the technical performance of new technical ideas to consideration of their cost and public acceptance, especially when the policy framework that can so seriously affect the success of particular technologies is so uncertain.

Rather than take action that depends on prediction of future market conditions, especially prices, or extend endless subsidies to particular technologies that are "too good to fail," the DOE should focus on demonstrating the technical performance, economics, and environmental effects of alternative technologies and thus create options for the private sector to choose or not.

The DOE has not been successful in this vital demonstration activity. The lack of success means that the pace at which new energy technology will be deployed will be slower than it might have been, which in turn means that the economy will be paying more for energy services than necessary. In fact, energy innovation is constrained not by an absence of new ideas but by the absence of early examples of successful implementation. Government action to improve the situation requires both more resources and a willingness to change the conventional approach to government's support for energy technology commercialization.

6

RECOMMENDATIONS

Given the importance of energy policy for the future welfare of Americans and others, and the lack of a reasonable policy over the past three decades, the question is what to do and how to do it.

Perhaps nothing can be done. Energy issues are so complex and involve so many conflicting interests that the U.S. political system may be incapable of finding a workable compromise that retains substance sufficient to make progress toward solving key issues. The linkage of domestic and international aspects, the tendency of political leaders to express idealistic goals rather than do the hard work of crafting policy based on analysis, the challenge of encouraging energy innovation in the face of technical uncertainty, and special interests are certainly difficult hurdles to overcome.

On the other hand, the U.S. government has shown the ability throughout its history to adapt to changes in economic, political, and technical realities. Two world wars caused the United States to shift fundamentally in its willingness to become involved and lead in international affairs. The Great Depression led to expanded federal authority and responsibility for the economic security of Americans. The development of the atomic bomb in World War II established the precedent for the federal government to be centrally involved in the technology developments that intrinsically affect U.S. national security and economic growth. And over the decades we have made some limited progress on energy matters: energy prices were deregulated by the historic legislation introduced by President Carter, effective environmental legislation has put into place workable processes to balance energy production and environmental protection, and export controls and other nonproliferation policies have slowed the spread of sensitive technology from commercial nuclear power fuel cycle operations.

So change in how the U.S. government operates is possible, but not easy. Often the motivation for change arises from a crisis and the public's expectation for action. The change, of course, should be genuine and affect underlying causes. Change in response to crisis all too often leads the political system to offer cosmetic organizational changes that are only an illusion of response. There are no better examples than the reorganization of the intelligence community and the establishment of a Department of Homeland Security, in response to the 9-11 terrorist attack on the United States in 2001. These were organizational changes that obscured rather than resolved the root causes of the inadequate cooperation between intelligence and law enforcement.

Many believe that ideological differences in Washington preclude action on energy or any other serious matter on the country's agenda. No substantive change may be possible, no matter how well justified by analysis. I do not dispute that the partisan climate is worse today than at any time during the past half century. Yet during the federalist period, arguably the most contentious time in domestic politics, significant change occurred—for example, legislators agreed on the Louisiana Purchase and the establishment of a national bank. In any case, it may be a long time until there is a bipartisan climate in Washington conducive to change.

Energy is only one of many important issues on the national agenda; others include climate, health care, economic recovery, the fiscal deficit, security issues (including terrorism), the risk of nuclear proliferation, and the conflicts in Afghanistan and Iraq. There are two implications that follow from this crowded agenda. First, it is unlikely that energy will be seen as sufficiently more important than these other issues to justify giving it priority in a political reform effort. Second, it is doubtful, in my view, that there is much chance of crafting a coalition for general reform of government processes to improve the country's ability to deal with all these issues. The issues are very different, and they involve different groups as winners and losers in any restructuring.

Thus, it is difficult to be optimistic about the prospects for change and tempting to conclude that the only option is to work within the existing system despite its flaws and limited opportunities for new policy approaches. This view prevails among the academic policy community, who would term it "realistic and pragmatic." My concern is that it has been shown that the realistic and pragmatic approach does not to lead to substantive progress on energy issues and amounts to wasting time and effort in return for just the intellectual satisfaction of working on an important public issue.

I propose a different approach: begin by identifying changes in the energy policymaking process that, if implemented, would enable greater and more rapid progress. Having clearly in mind what changes should be made is a necessary step toward making the changes. It is to be hoped that debate about the recommendations will influence the thinking of policy leaders on the urgent need for fundamental change to replace the frustrating and inconclusive energy policy debates of the day. As I indicated in the Introduction, the changes I propose are unlikely to be implemented soon.

Five Proposals

With one exception, the changes I propose are to the policy-making process, not bureaucratic reorganization or creation of new agencies. My mentor and the DOE's first secretary, James Schlesinger, referred to a diagram of a reorganization proposal as "different tree, same monkeys," because all too often the reorganization is cosmetic and does not confront the underlying problem. I offer five proposals intended to redress the shortcomings discussed earlier in this book.

1. Better integrate domestic and international energy policy. As discussed in Chapter 3, the central, but not exclusive, energy security concern is oil import dependence. It is not possible for the United States to achieve oil import independence. And even if it were, our allies would remain dependent on oil imports, so the United States would still face security concerns, albeit indirectly. Those responsible for energy policy at the White House must improve its focus on the linkage between the domestic and the international aspects of energy policy. Every decision on a domestic energy matter has a foreign policy implication, and many foreign policy decisions can have a major impact on domestic energy matters. For example, the U.S. reaction to the massive BP oil spill at the Macondo platform in the Gulf of Mexico will influence the measures other countries adopt to regulate deep water offshore oil and gas drilling.

The United States has not effectively established linkage between domestic policy and foreign policy. Inevitably such linkage must take place in the executive office of the president, where the equities of various agencies—State, Energy, Defense, EPA, and others—are debated, weighed, and decided. There is no single mechanism that ensures effective integration. Many different arrangements are possible, such as creating an assistant to the president for energy matters, or vesting responsibility in the National Security Council or the National Economic Council. There are pros and cons for each of these arrangements, and every president should be expected to choose an approach that meets the particular circumstances and personalities of his or her administration. The Obama administration in its early months appeared to choose all of these mechanisms, and their efficacy is as yet unknown.

Here are two examples of the linkage between energy and security perspectives that require policy integration. First, significant growth in demand for oil is coming from the larger emerging developing economies, notably China, India, Indonesia, Brazil, and Mexico. U.S. diplomacy needs to engage these countries on (1) establishing the joint benefit from transparent, open oil markets rather than bilateral state-to-state producer–supplier agreements; (2) planning for new oil and gas pipeline routes from Central Asia and the Caspian region; and (3) initiatives to reduce the vulnerability of the international energy infrastructure to natural disasters and possible attacks by terrorists groups.

The second example is the continuing dependence of oil-importing countries on oil production from the Persian Gulf, in particular Iran, Iraq, Saudi Arabia, and Kuwait. Persian Gulf nations are anticipated to furnish about 50% of OPEC oil production in 2030, which is expected to be about 40% of world oil production.[1] Inevitably the geopolitical stability of this region is important to the United States and other oil-importing countries, and this concern will influence the options the United States and other countries will be willing to consider in advancing other important foreign policy interests. The most vivid example is Iran, where the advantages of greater and reliable Iranian oil production for world markets must be balanced against the objective of blocking Iran's path to a bomb. A strong U.S.

military and diplomatic presence in the region is important for both energy and broader national security interests. Energy and security interests are inexorably linked in Persian Gulf. With the anticipated increase in unconventional natural gas supplies in the United States and elsewhere in the world, the United States will need to adopt a policy about North American exports of natural gas, which has important domestic and international ramifications.[2]

2. Establish a single joint energy and environment committee in the U.S. Congress similar to the highly successful Joint Committee on Atomic Energy or Joint Economic Committee of the past. In the U.S. political system, any change in policy or program requires the support and approval of Congress. I have indicated at several points the significant responsibility that Congress bears for failed energy policy: Congress has an inevitable tendency to put local interest above national interest in shaping energy policies and programs. Chapter 2 described the inability of Congress to pass constructive climate legislation.

A new way of doing business in the Congress could give a sense of urgency to energy issues, greatly speed the legislative process, and dampen the influence of local interests that can at present, for example, block the siting of nuclear waste disposal facilities; inflate subsidies for gasohol, as discussed earlier; and prolong the inability to reach agreement on a carbon policy. A change in committee structure does not ensure that the underlying differences will disappear, but a joint committee may be able to give greater weight to a national, as opposed to a local, perspective. I believe that a significant change is needed in how Congress organizes to deal with energy issues, if progress is to be made.

3. Rebalance federal and state authority over energy issues, such as electrical transmission and pricing, and regulations to reduce emissions through renewable portfolio or fuel standards. As discussed in Chapters 1 and 2, the absence of action on energy issues over the past two decades has led to a consistent trend shifting responsibility away from the federal government (with a national perspective) to state governments (with a regional perspective). Inevitably a state or regional approach leads to policies that are more responsive to local interests. Moreover, Congress has been reluctant to circumscribe local and regional authority even in situations where national preemption makes sense—for example, granting the Federal Energy Regulatory Commission (FERC) the authority to site interstate electrical transmission lines and to set national procedures for managing the pricing and dispatch of electricity on the grid.

There is no better example of how the absence of federal action on energy issues encourages states to take the initiative than climate change and renewable energy. Absent a national policy to set a price on carbon emissions, individual states such as California, Texas, and Massachusetts have adopted local regulations, including Renewal Portfolio Standards, Renewable Fuel Standards, and, in the case of California, ambitious and complex programs for zero emission vehicles.[3] Many states also offer tax benefits for energy conservation or renewable energy projects and support RD&D in local universities and industry.

Granting that the states have good intentions in taking these energy initiatives, the trend cannot be considered constructive. Efficient energy policies require a national perspective (some would argue, a global perspective), which is unachievable through a patchwork of local policies. Different policies among states can create perverse economic incentives and certainly problems for harmonizing regulation if and when national polices are put in place. The United States would be aghast if the large, rapidly growing, emerging countries such as China and India were to organize their energy policy efforts in this way. All the OECD countries have a more centralized energy policy apparatus than does the United States.

The debate about the split in authority for energy and environmental regulation has gone on for many years. Much can be said on both sides of this argument, especially when it comes to inspection and enforcement—for example, for clean air and clean water. However, energy problems such as regulating greenhouse gas emissions, disposing of nuclear waste, and paying for R&D are intrinsically national in character. It is time for Congress to embrace the concept that certain energy issues demand a national approach and authorize federal preemption for policies that govern those issues.

4. Plans and numbers. The executive branch should annually prepare a five-year program plan of energy and related environmental activities undertaken by all federal agencies. Congress should approve this plan as the authorization basis for annual agency budgets. Congress should adopt a rule that all legislation submitted for consideration, either by the executive branch or by members of Congress, should be supported by analysis. The analysis could be prepared by an executive branch agency (endorsed by the Office of Management and Budget) or by the Congressional Research Service, or the Congressional Budget Office.

The requirement for an analytic support for all energy legislative proposals does not ensure sound policy, but it will prevent, or at least constrain, policymakers from setting policy goals before any quantitative analysis is even considered. Barefoot proposals that either are impossible to achieve or simply state a desirable goal without definition or analytic support (for instance, "green jobs") will be more difficult to sustain in debate.

A comprehensive national energy plan is an absolute prerequisite, for the executive branch, Congress, and the public to be able to monitor progress in meeting specified objectives within cost and on schedule.

The DOE's EIA National Energy Modeling System (NEMS) is a necessary piece, but it is based primarily on extrapolation of historical data.[4] It must be upgraded to deal with a wider range of scenarios and include modern tools to evaluate and assess risk. The government planning effort should provide for continuous interaction with academic and industrial planning organizations in order to ensure that it benefits from the most advanced and adventuresome thinking in the country.

In November 2010, the President's Council of Advisors on Science and Technology released a report, *Government-Wide Energy Technology Strategic Planning and New Investments in Energy Research, Development, and Deployment,* that recommends a White House–sponsored Quadrennial Energy Review (QER).[5] The QER would establish national goals, the multiyear resources requirements needed to support policy options, as well as incentives and regulations to facilitate desired activity. The Executive Office of the President would lead the QER, with the Department of Energy providing a secretariat. Here is a reasonable starting point for a process to maintain an up-to-date multiyear energy plan for the country. Of course, it is essential that the QER be supported by serious quantitative analysis—not a strong suit in any White House—thus making shared responsibilities between the Department of Energy and the Executive Office of the President absolutely vital. The White House brings a desirable presidential and multi-agency perspective but also a tendency to announce ambitious goals in absence of any analysis or thought about the feasibility of achieving the goal. On the other hand, the Department of Energy has necessary domain knowledge, but also an organization's natural resistance to change and preference for existing practice. To be successful the QER mechanism would need to combine the strengths of both entities. Some readers will observe that multi-year resource plans for activities that span several agencies would be useful for matters other than energy; quite so.

5. *Strengthen the federal government's efforts to encourage energy innovation.* As discussed in Chapter 5 and elsewhere in this book, market-driven technology innovation is key to steady progress in deploying cost-effective solutions to energy supply and demand needs. I have identified three critical steps for the DOE:

(a) Continue the DOE's support of the early-stage innovation process. Early R&D creates new knowledge and options for solving energy problems. The government has a special role in supporting this early discovery phase of innovation, and in the United States, the government has done an admirable job of fostering basic technology in the areas of health, defense, and energy. The DOE's record in early-stage R&D is strong—historically, through the Office of Basic Energy Sciences supporting research in universities, government laboratories, and industry in a traditional manner; more recently, through innovative new approaches such as ARPA-E. Only a small fraction of these ideas will have dramatic effect on how we use energy, but those that do succeed will benefit the United States both directly in how we use energy and indirectly in the prospect of exporting new energy products and services.

A great deal more could be accomplished if the technology base efforts of the various agencies—the DOE, the National Science Foundation, the Departments of Agriculture, Commerce, Defense, and others—were rationalized for key technology areas such as catalysis, batteries, photochemistry, biofuels, new materials, and improving energy efficiency. The Office of Science and Technology Policy (OSTP) and the Office of Management and Budget should play a strong role in integrating energy technology base programs across the federal government.

(b) The key transition from laboratory scale research to development and possible demonstration requires discipline management based on explicit analysis. Modeling and simulation tools that assess the technical and economic consequences of various system choices are essential for this analysis. The federal government now does not have adequate modeling and simulation capability. At the level of energy supply and use, the DOE should dramatically increase its capability to model and simulate trade-offs between technical performance, cost, and environmental attributes. This capability, which is much cheaper than demonstration projects, is especially needed for nuclear power, clean coal, smart grid, and improved energy efficiency. The modeling and simulation activity should include engineering data, and should also include cooperation and review with industry and private groups that have regional and field experience. The development of advanced computational tools that permit the modeling and simulation of the technical performance and economic optimization of energy alternatives has the potential for more rapid and efficient identification of desirable development paths.

(c) Congress should create an Energy Technology Corporation (ETC) to manage demonstration projects. As discussed in Chapter 5, one of the recurring weaknesses in federal RD&D is the demonstration phase. Too often this expensive stage in the energy innovation process is carried out in a manner that provides little useful information to the private sector. The ETC would be a new, semipublic organization, governed by an independent board of individuals nominated by the president and confirmed by the Senate, for the purpose of financing selected large-scale demonstration projects in a manner that is commercially credible. As explained in Chapter 5, the ETC differs from an industry-managed consortium, a green development bank, and the ill-fated U.S. Synthetic Fuels Corporation because of its focus on demonstrating the technical and economic readiness of important new energy technologies.

These five proposals illustrate the kind of change needed to improve the way the United States sets energy policy. These proposals are not necessary and sufficient conditions to move from a situation where the country is unable to adopt sensible energy policy to a utopian situation where optimal energy policy is adopted by acclamation. Energy policy will always be difficult, because different policy choices have outcomes that impact groups differently. We should seek and expect a policy process that clearly frames choices in a specific if-then format: "If this policy is adopted, then these consequences follow." At present, those who advocate policy often do not feel obligated to suggest a credible path to accomplish the objectives or to offer an accurate assessment of likely results.

Absent changes of the character I propose, I suggest that sensible U.S. energy policy will not happen. The United States will not adopt effective climate change legislation, make progress on reducing oil import dependence, or begin the long transition from an economy based on fossil fuels to one reliant on renewable energy technologies and nuclear power. The result will be that generations of Americans will face energy costs higher than necessary and endure more serious social disruption, including, perhaps, conflict that results because critical energy issues have not been resolved.

Getting Started

The five-point action plan will not be adopted quickly, so it is worthwhile to consider a staged approach to the desired objective. The first stage most certainly should be devoted to laying out the case for change to a wide audience: leaders of the political, industrial, and academic communities as well as the public. The logic of the case is simple:

- Energy is important to the country's welfare.
- We have been unsuccessful in agreeing on a long-range energy plan for the country that will meet the key objectives of (1) reducing the risks of climate change; (2) improving energy security, including beginning to reduce dependence on imported oil; and (3) beginning the long transition from an economy based on fossil fuels to an economy based on renewable sources of energy.
- A change in the approach to how energy policy is addressed by our political system is a necessary prerequisite for progress.

My hope is for a debate about the need for a change and the means for accomplishing change. Colloquia at universities and research centers would begin to address the shortcomings of energy policymaking in the United States rather than the pros and cons of the most recent version of energy legislation that is inadequate to meet the energy challenges listed above. Success in the first stage would be greater public awareness, measured by a variety of means: polls, articles, and speeches by notables affirming that something is wrong with the process of policymaking rather than with the content of a particular bill or executive branch initiative.

The second stage is for the executive branch to put into place those initiatives that can be done without legislative authority. This, I admit, is an expedient approach, reflecting the greater difficulty in achieving change that involves both the executive branch and the legislative branch. Moreover, I believe that strong and determined executive branch commitment is needed if fundamental political process change is to happen. So if the executive branch is unable or unwilling to take a new approach, there is little possibility that the legislative branch will do so.

The executive branch can accomplish some of my proposals without any new legislative authority or significant perturbation to legislative process. These actions could be the basis of a new energy and climate legislative initiative, solidly based on quantitative analysis and a multiyear integrated program. It would be the president's program and a strong starting point for legislative debate—a very different approach to energy legislation than we have seen in the past decade.

Success in stage two would make it easier in stage three to accomplish passage of legislation that realigns authorities. Passage of sound energy legislation implies congressional recognition of steps that are needed to protect the country's energy future. So perhaps it is not overly optimistic to believe that Congress would be prepared to support two measures: rebalancing federal and state energy authorities, and creating an Energy Technology Corporation (ETC) to manage demonstration projects, on the grounds that these measures are essential to implementing comprehensive energy and climate legislation.

The final proposal, to establish a single joint energy and environment committee in the U.S. Congress, would be a dramatic step, but probably a bridge too far.

Readers of this book should reach several conclusions: First, energy is important to the future welfare of the United States, and failure to adopt sound energy policies today almost certainly will mean greater cost and hardship for future generations. Second, adopting sound energy policies requires significant change in how Congress and the executive branch manage the legislative process. Third, sound energy policy should be based on systematic analysis that requires quantitative thinking relying on real data to ensure that scarce resources are not wasted. Fourth, the continuing challenge is to harmonize analysis that is inherently uncertain, based inevitably on limited methodology and data, and often reflecting organizational bias against change, with a political process that seizes a popular vision of a short time horizon limited to the next election and a passionate desire to defend the interests of constituencies.[6] The struggle to improve rational energy policymaking is ongoing, dependent on trained and dedicated individuals who are intent on sharpening the judgment of decision makers rather than using analysis to defend a preconceived point of view.

NOTES

INDEX

NOTES

1. The Failure of U.S. Energy Policy

1. For the Clean Air Act, see http://www.epa.gov/air/caa/.

2. http://www.epa.gov/air/urbanair/.

3. *Restoring the Quality of the Environment: Report of the Environmental Pollution Panel of the President's Scientific Advisory Committee* (The White House, November 1965). See the report of the Subcommittee on Carbon Dioxide in the Atmosphere, appendix Y4.

4. The major anthropogenic GHGs are carbon dioxide (CO_2), methane (CH_4), nitrous oxide (N_2O), and fluorinated hydrocarbons. The relative importance of these species depends on both the strength of absorption and their lifetimes in the atmosphere. In 2008 the relative emissions in terms of CO_2 equivalents were 5,921 metric tonnes (MT) CO_2, 567 MT CH_4, 318 MT N_2O, and 150 MT fluorinated hydrocarbons, for a total of 6,016 MT. About 40% of the CO_2 emissions come from coal used in electricity generation. Water vapor is the most significant GHG, but its presence in the atmosphere comes from natural processes. http://www.epa.gov/climatechange/emissions/usinventoryreport.htm.

5. The IPCC work is found at http://www.ipcc.ch/publications_and_data/publications_and_data.htm.

6. For an exciting and discerning account of the meeting and U.S. policy leading up to it, see William J. Antholis and Strobe Talbott, *Fast Forward: Ethics and Politics in the Age of Global Warming* (Brookings Institution Press, 2010).

7. President Richard Nixon, State of the Union, January 30, 1974.

8. EIA, *Annual Energy Outlook* (2009), table 1, reference case.

9. President Jimmy Carter, public address from the White House, June 20, 1979.

10. President George W. Bush, State of the Union, January 31, 2006.

11. I rely heavily on the MIT Emissions Prediction and Policy Analysis (EPPA) model, described at http://globalchange.mit.edu/igsm/eppa.html. There are other modeling efforts. For example, the EPA uses a number of climate models, described at http://www.epa.gov/climatechange/economics/modeling.html.

12. Richard Lester and Ashley Finan, *Quantifying the Impact of Proposed Carbon Emission Reductions on the U.S. Energy Infrastructure,* MIT Industrial Performance–Energy Innovation Working Paper 09-006, October 2009, available at http://web.mit.edu/ipc/publications/pdf/09-006.pdf.

13. Toby Bolsen and Fay Lomax Cook, "Public Opinion Trends on Energy Policy, 1974–2006," *Public Opinion Quarterly* 72, no. 364 (2008).

14. Democratic administrations have generally been opposed to expanded domestic oil and gas drilling, especially in Alaska and offshore. It is noteworthy that President Obama announced on March 31, 2010, the administration's intention to expand exploration and production of oil and gas on the U.S. outer continental shelf. See http://www.whitehouse.gov/the-press-office/obama-administration-announces-comprehensive-strategy-energy-security. Although the increase in domestic production will not displace a significant fraction of imports, this action will have a positive effect on foreign producers because it shows that the United States is willing to increase domestic production as they exhort foreign producers to do so.

15. Eugene A. Rosa and Riley E. Dunlap, "Poll Trends: Nuclear Power—Three Decades of Public Opinion," *Public Opinion Quarterly* 58, no. 295 (1994). See also MIT, *The Future of Nuclear Power* (2003), chap. 9, available at web.mit.edu/nuclearpower.

16. Stephen Ansolbahere and David M. Konisky, "Public Attitudes toward Construction of New Power Plants," *Public Opinion Quarterly* 73, no. 566 (2009).

17. MIT, *The Future of Coal* (2007), chap. 7, available at web.mit.edu/coal.

18. March 10, 2010, Gallup Poll report, at http://www.gallup.com/poll/126560/Americans-Global-Warming-Concerns-Continue-Drop.aspx.

19. Howard Herzog et al., *2009 Survey of Public Attitudes on Energy and the Environment*, at http://sequestration.mit.edu/research/survey2009.html.

20. Several members of the Harvard Kennedy School faculty have made important contributions to proliferation policy: Professors Doty, Nye, Allison, Carnesale, Carter, Holdren, and Bunn.

21. See A. D. Ellerman, R. Schmalensee, E. Bailey, P. Joskow, and J.-P. Montero, *Markets for Clean Air: The U.S. Acid Rain Program* (Cambridge University Press, 2000).

22. The DOE's Energy Information Administration maintains the National Energy Modeling System (NEMS) for the purpose of projecting energy supply and use under three different scenarios. The NEMS system is very useful, but it does not use engineering data or pretend to explore different energy development trajectories—for example, different nuclear fuel cycles.

23. President Carter's televised speech, "The President's Proposed Energy Policy," April 18, 1977.

24. Ibid. President Carter's proposals included establishing a new Department of Energy. Available at http://www.pbs.org/wgbh/amex/carter/filmmore/ps_energy.html.

25. President Carter signed the DOE Organization Act on August 4, 1977, U.S. 91 Stat. 569. The department began operation on October 1, 1977. It is worthwhile to reflect on Section 7112, the Congressional Declaration of Purpose (U.S. Code 42.84.7112).

The Congress therefore declares that the establishment of a Department of Energy is in the public interest and will promote the general welfare by assuring coordinated and effective administration of Federal energy policy and programs. It is the purpose of this chapter:

(1) To establish a Department of Energy in the executive branch.

(2) To achieve, through the Department, effective management of energy functions of the Federal Government, including consultation with the heads of other Federal departments and agencies in order to encourage them to establish and observe policies consistent with a coordinated energy policy, and to promote maximum possible energy conservation measures in connection with the activities within their respective jurisdictions.

(3) To provide for a mechanism through which a coordinated national energy policy can be formulated and implemented to deal with the short-, mid- and long-term energy problems of the Nation; and to develop plans and programs for dealing with domestic energy production and import shortages.

(4) To create and implement a comprehensive energy conservation strategy that will receive the highest priority in the national energy program.

(5) To carry out the planning, coordination, support, and management of a balanced and comprehensive energy research and development program, including—
   (A) assessing the requirements for energy research and development;
   (B) developing priorities necessary to meet those requirements;
   (C) undertaking programs for the optimal development of the various forms of energy production and conservation; and
   (D) disseminating information resulting from such programs, including disseminating information on the commercial feasibility and use of energy from fossil, nuclear, solar, geothermal, and other energy technologies.

(6) To place major emphasis on the development and commercial use of solar, geothermal, recycling and other technologies utilizing renewable energy resources.

(7) To continue and improve the effectiveness and objectivity of a central energy data collection and analysis program within the Department.

(8) To facilitate establishment of an effective strategy for distributing and allocating fuels in periods of short supply and to provide for the administration of a national energy supply reserve.

(9) To promote the interests of consumers through the provision of an adequate and reliable supply of energy at the lowest reasonable cost.

(10) To establish and implement through the Department, in coordination with the Secretaries of State, Treasury, and Defense, policies regarding international energy issues that have a direct impact on research, development, utilization, supply, and conservation of energy in the United States and to undertake activities involving the integration of domestic and foreign policy relating to energy, including provision of independent technical advice to the President on international negotiations involving energy resources, energy technologies, or nuclear weapons issues, except that the Secretary of State shall continue to exercise primary authority for the conduct of foreign policy relating to energy and nuclear nonproliferation, pursuant to policy guidelines established by the President.

(11) To provide for the cooperation of Federal, State, and local governments in the development and implementation of national energy policies and programs.

(12) To foster and assure competition among parties engaged in the supply of energy and fuels.

(13) To assure incorporation of national environmental protection goals in the formulation and implementation of energy programs, and to advance the goals of restoring, protecting, and enhancing environmental quality, and assuring public health and safety.

(14) To assure, to the maximum extent practicable, that the productive capacity of private enterprise shall be utilized in the development and achievement of the policies and purposes of this chapter.

(15) To provide for, encourage, and assist public participation in the development and enforcement of national energy programs.

(16) To create an awareness of, and responsibility for, the fuel and energy needs of rural and urban residents as such needs pertain to home heating and cooling, transportation, agricultural production, electrical generation, conservation, and research and development.

(17) To foster insofar as possible the continued good health of the Nation's small business firms, public utility districts, municipal utilities, and private cooperatives involved in energy production, transportation, research, development, demonstration, marketing, and merchandising.

(18) To provide for the administration of the functions of the Energy Research and Development Administration related to nuclear weapons and national security, which are transferred to the Department by this chapter.

(19) To ensure that the Department can continue current support of mathematics, science, and engineering education programs by using the personnel, facilities, equipment, and resources of its laboratories and by working with State and local education agencies, institutions of higher education, and business and industry. The Department's involvement in mathematics, science, and engineering education should be consistent with its main mission and should be coordinated with all Federal efforts in mathematics, science, and engineering education, especially with the Department of Education and the National Science Foundation (which have the primary Federal responsibility for mathematics, science, and engineering education).

26. In contrast to the 1978 National Energy Plan I, the 2001 National Energy Plan consists mostly of philosophy and statements of future energy outcomes without setting priorities or offering quantitative projections or specific policies. Available at http://www.wtrg .com/EnergyReport/National-Energy-Policy.pdf.

27. The *Annual Energy Outlook (AEO)* is found at http://www.eia .doe.gov/oiaf/aeo/, and the *International Energy Outlook (IEO)* is found at http://www.eia.doe.gov/oiaf/ieo/.

28. The National Energy Act of 1978 deregulated natural gas prices, prohibited use of natural gas in major fuel-burning installations (e.g., power plants), and required electric utilities to accept electricity from qualified cogeneration facilities. The Energy Security Act of 1979 established the windfall profits tax on deregulated oil and the synthetic fuels corporation.

29. http://media.washingtonpost.com/wp-srv/nation/pdf/Power ActDraft_051110.pdf.

30. See the California Air Resources Board website: http://www .arb.ca.gov/cc/scopingplan/scopingplan.htm.

31. The RGGI initiative is described at http://www.rggi.org/home.

32. The EPA endangerment finding is described at http://www.epa .gov/climatechange/endangerment.html.

2. Energy and Climate Change

1. The Kyoto Protocol designates six most important GHGs: carbon dioxide, methane, nitrous oxide, hydrofluorocarbons, perfluorocarbons, and sulfur hexafluoride. The warming effect depends upon these molecules' abundance, their relative absorption of infrared radiation, and their lifetime in the atmosphere. The principal sources of these emissions are fossil fuel power production, industrial and agricultural processes, and transportation. Water vapor absorbs most of the infrared radiation from the sun that passes through the atmosphere, but it does not come from anthropogenic emissions.

2. The most compelling account of the U.S. history of climate change is in William Antholis and Strobe Talbott, *Fast Forward: Ethics*

and Politics in the Age of Global Warming (Brookings Institution Press, 2010).

3. http://unfccc.int/kyoto_protocol/items/2830.php.

4. http://unfccc.int/meetings/cop_15/items/5257.php. The Copenhagen accord is available at http://unfccc.int/resource/docs/2009/cop15/eng/l07.pdf.

5. See the IPCC website, www.ipcc.ch.

6. For the U.N. Framework Convention on Climate Change, see their website: http://unfccc.int/2860.php. The Kyoto Protocol was adopted at COP 3 in 1997.

7. Y. Kaya, "Impact of Carbon Dioxide Emission Control on GNP Growth: Interpretation of Proposed Scenarios," paper presented to the IPCC Energy and Industry Subgroup, Response Strategies Working Group, Paris, 1990 (mimeo).

8. For the mathematically inclined, the identity result is found from the product identity $C = (C/E)(E/Y)(Y)$ and taking small differentials of each quantity and then dividing by the product.

9. EIA, *International Energy Outlook* (2010), highlights at http://www.eia.doe.gov/oiaf/ieo/. Kaya data in appendix J.

10. EIA, *Annual Energy Review* (2008).

11. Available at http://english.gov.cn/2009–11/27/content_1474764.htm.

12. A number of international trade effects need to be evaluated to determine what the net effect is. For an interesting recent attempt to do so for the United Kingdom, see G. Baiocchi and J. C. Minx, "Understanding Changes in the U.K.'s CO_2 Emissions: A Global Perspective," *Environmental Science & Technology* 44, no. 4 (2010): 1177.

13. It is possible to imagine other mechanisms that could remove the incentive for energy intensive industries to move to locations with lower emission charges—for example, an international agreement on an industry emission charge, regardless of location. The agreement would need to specify how the revenue would be distributed and a mechanism to ensure compliance.

14. EIA, *International Energy Outlook* (2009), reference case projections.

15. Ibid. (2010), highlights at http://www.eia.doe.gov/oiaf/ieo/.

16. The DOE uses the National Energy Modeling System (NEMS), http://www.eia.doe.gov/oiaf/aeo/overview/, which was developed by the EIA. The EPA uses the Applied Dynamic Analysis of the Global Economy (ADAGE) model and the Intertemporal General Equilibrium Model (IGEM) for climate economic modeling, http://www.epa.gov/climatechange/economics/modeling.html. There are many academic and NGO models, such as the MIT EPPA model described in this chapter, that are frequently discussed at the respected Stanford Energy Modeling Forum, http://emf.stanford.edu/research/emf20/. Models address all aspects of energy production, transportation, and use, including the influence of policy. There are also models to assist decision makers, such as MIT's C-ROADS simulation model; see http://jsterman.scripts.mit.edu/docs/Sawin-2009-CRoads.pdf.

17. Sergey Paltsev, John M. Reilly, Henry D. Jacoby, and Jennifer F. Morris, *The Cost of Climate Policy in the United States,* MIT Joint Program on the Science and Policy of Global Change, Report #173, April 2009.

18. Table N.1 gives the data on which Figure 2.2 is based. The midcentury GHG emissions are not 20% of the 2005 values, because the cap-and-trade mechanism importantly permits "borrowing" of emission allowances.

19. See MIT, *The Future of Coal* (2007), chap. 3, available at http://web.mit.edu/coal/.

20. EIA, *Annual Energy Outlook* (2009), table A-18, reference case.

21. Recall that air is about 80% nitrogen and 20% oxygen.

22. See G. Baiocchi and J. C. Minx, "U.K.'s CO_2 Emissions."

23. M. Hamilton and H. Herzog, *Cost Update for the Future of Coal,* MIT Energy Initiative, July 2008. See also M. Hamilton, H. J. Herzog, and J. E. Parson, *Cost and U.S. Public Policy for New Coal Power Plants with Carbon Capture and Sequestration, Energy Procedia* 1, no. 4487 (2009). Recent EIA estimates for capital costs are considerably higher: PC w/o CCS $2,844 per kWe and IGCC w/CCS $5348 per kWe. See http://www.eia.gov/oiaf/beck_plantcosts/index.html.

Table N.1. Data used for Figure 2.2

Exajoules	Scenario #1 Business as usual	Scenario #2 Reference response	Scenario #3 Nuclear advantage	Scenario #4 CCS advantage	Scenario #5 Renewable advantage
Renewable	5.8	9.6	9.1	9.8	16.7
Biomass liquids	3.1	11	3.1	12.7	19
Nuclear	8	22.1	41	8.3	8.4
Natural gas	23.7	18.9	14	23.3	25.2
Petroleum	62.5	31	39.9	29.5	22.4
Coal w/o CCS	39.7	9.4	2	22.9	2.1
Coal w/CCS	0	7–9	0	21.2	0
Energy demand loss	0	32.8	33.5	15	48.9
GHG emissions Gtonnes	10.74	3.38	3.79	3.36	344
CO_2 equivalent					
Prices ($2005)					
Oil $/b	154	139	141	139	139
Coal $/MMBTU	1.80	1.46	1.39	1.56	145
Natural gas $/MCF	18	9	8	10	9
Electricity ¢/kWe-h	13	19	17	19	19
CO_2 equiv. price $/tonne	0	229	190	235	266

Gtonnes = 10^9 metric tons, MCF = thousand cubic feet, MMBTU = million British Thermal Units.

24. DOE estimates from the National Energy Technology Laboratory (NETL) are summarized at http://www.netl.doe.gov/publications/factsheets/program/Prog065.pdf.

25. See the MIT Energy Initiative Symposium report, *Retrofitting of Coal-Fired Power Plants for CO$_2$ Emissions Reductions* (March 2009), available at http://web.mit.edu/mitei/docs/reports/coal-paper.pdf.

26. A recent issue of *Science* magazine devoted to carbon sequestration describes several options: see *Science* 325 (September 25, 2009).

27. See, for example, the IPCC report *Carbon Dioxide Capture and Storage* (2005), which includes a summary for policymakers. Available at http://www.iph/pdf/special-reports/srccs/srccs_wholereport.pdf. Also see Baiocchi and Minx, "U.K.'s CO$_2$ Emissions," chap. 5.

28. Schlumberger maintains a website that describes a rigorous carbon sequestration system. (Disclosure: I have been a director and advisor to this company for more than twenty years.) http://www.slb.com/content/services/additional/carbon/index.asp.

29. World Resources Institute, *Guidelines for Carbon Dioxide Capture, Transport, and Storage* (2008). Available at http://pdf.wri.org/ccs_guidelines.pdf.

30. Carbon capture and sequestration technologies @ MIT, http://sequestration.mit.edu/.

31. International Energy Agency, *CO$_2$ Capture and Storage* (2008), available at http://www.iea.orgr/textbase/nppdf/free/2008/CCS_2008.pdf.

32. The DOE carbon sequestration budget from FY97 to FY2010 can be found at http://www.netl.doe.gov/technologies/carbon_seq/refshelf/project%20portfolio/2009/Overview/FY%202009%20Budget%20.

33. The DOE carbon sequestration technology and program plan as of 2007 describes in detail the activities under way at that point. Available at http://www.netl.doe.gov/technologies/carbon_seq/refshelf/project%20portfolio/2007/2007Roadmap.pdf.

34. For the regional partnerships program, see http://www.netl.doe.gov/technologies/carbon_seq/partnerships/partnerships.html.

35. http://www.adm.com/en-US/responsibility/Documents/Carbon-Sequestration-Brochure.pdf.

36. M. F. Pollak and E. J. Wilson, "Regulating Geologic Sequestration in the U.S.: Early Rules Take Divergent Approaches," *Environmental Science & Technology* 43, no. 3035 (2009).

37. See MIT, *The Future of Coal* (2007), chap. 3, available at http://web.mit.edu/coal/.

38. For example, Daniel P. Schrag, "Making Carbon Capture and Storage Work," in *Acting in Time on Energy Policy,* ed. Kelly Gallagher (Brookings Institution Press, 2009), p. 39; Schrag, "Preparing to Capture Carbon," *Science* 315, no. 5813 (2007): 812.

39. The DOE announcement is found at http://www.fossil.energy.gov/news/techlines/2010/10033-Secretary_Chu_Announces_FutureGen_.html. Also see http://www.fossil.energy.gov/news/techlines/2010/10038-DOE%2C_FutureGen_Alliance_Discuss_Ne.html.

40. DOE press release: http://www.energy.gov/8356.htm.

41. There is an extensive literature on cap-and-trade. A good starting point is A. Denny Ellerman, M. D. Webster, J. Parsons, H. D. Jacoby, and M. McGuinness, *Cap and Trade: Contributions to the Design of a U.S. Greenhouse Gas Program,* MIT Center for Energy and Environmental Policy Research (2008). Available at http://web.mit.edu/ceepr/www/publications/DDCF.pdf.

42. Remarks of President Obama at Carnegie Mellon University, June 2, 2010. Available at www.whitehouse.gov/the-press-office/remarks-president-economy-carnegie-mellon-university.

43. Environmental Protection Agency, *Analysis of H.R. 2454, the American Clean Energy and Security Act of 2009;* see http://www.epa.gov/climatechange/economics/economicanalyses.html#hr2454 (June 2009). Energy Information Administration, *Energy Market and Economic Impacts of H.R. 2454, the American Clean Energy and Security Act of 2009* (August 2009), http://www.eia.doe.gov/oiaf/servicerpt/hr2454/index.html.

44. National Commission on Energy Policy, *Domestic and International Offsets* (September 15, 2009), http://www.bipartisanpolicy

.org/sites/default/files/NCEP%20Domestic%20and%20International %20Offsetsformatted.pdf.

45. Ryan Lizza, "As the World Burns: How the Senate and the White House Missed Their Best Chance to Deal with Climate Change," *New Yorker,* October 11, 2010.

46. http://www.energy.gov/recovery/data.htm.

47. In April 2007, the U.S. Supreme Court ruled in the case *Massachusetts v. Environmental Protection Agency,* 549 U.S. 497 (2007), that the EPA violated the Clean Air Act by not regulating greenhouse gas emissions.

48. President Obama's speech can be found at http://www. whitehouse.gov/the-press-office/2011/03/30/remarks-president -americas-energy-security.

49. The Blueprint for a Secure Energy Future may be found at http://www.whitehouse.gov/sites/default/files/blueprint_secure_ energy_future.pdf.

50. There remain some optimists. I take seriously the views of Michael Spence and Rob Stavins. See Michael Spence, *Climate Change, Mitigation, and Developing Country Growth,* Commission on Growth and Development, Working Paper #64 (2009), http://www.growth-commission.org/storage/cgdev/documents/gcwp064812web.pdf; Robert N. Stavins, *The Path Forward for Climate Talks,* http://www .hks.harvard.edu/fs/rstavins/Forum/Column_35.pdf.

3. ENERGY SECURITY

1. A more complete presentation of my views on these issues can be found in John Deutch and James R. Schlesinger, co-chairs, *National Security Consequences of U.S. Oil Dependence,* Council of Foreign Affairs Task Force Report (October 2006), available at http:// www.cfr.org/publication/11683/national_security_consequences_of _us_oil_dependency.html?breadcrumb=%2Fissue%2F17%2Fenergy environment.

2. Olivier Blanchard and Marianna Riggi, *Why Are the 2000s So Different from the 1970s? A Structural Interpretation of Changes in the*

Macroeconomic Effects Of Oil Prices, NBER Working Paper Series #15467 (October 2009), available at http://www.nber.org/papers/w15467.

3. The Organization for Economic Cooperation and Development (OECD), established in 1961, consists of thirty countries in North America, and Japan and Korea in Asia.

4. For a paper that addresses the security implications of the evolving natural gas trade, see MIT, *The Future of Natural Gas* (2010), available at http://web.mit.edu/mitei/research/studies/naturalgas.html.

5. Table A.5, World's Liquid Consumption by Region, to 2035, Reference Case EIA International Energy Outlook, 2010. Available at http://www.eia.doe.gov/oiaf/ieo/ieorefcase.html.

6. President Jimmy Carter in his State of the Union address on January 23, 1980; http://www.jimmycarterlibrary.org/documents/speeches/su80jec.phtml. A longer excerpt from this address is worth reading:

> The region which is now threatened by Soviet troops in Afghanistan is of great strategic importance: It contains more than two-thirds of the world's exportable oil. The Soviet effort to dominate Afghanistan has brought Soviet military forces to within 300 miles of the Indian Ocean and close to the Straits of Hormuz, a waterway through which most of the world's oil must flow. The Soviet Union is now attempting to consolidate a strategic position, therefore, that poses a grave threat to the free movement of Middle East oil.
>
> This situation demands careful thought, steady nerves, and resolute action, not only for this year but for many years to come. It demands collective efforts to meet this new threat to security in the Persian Gulf and in Southwest Asia. It demands the participation of all those who rely on oil from the Middle East and who are concerned with global peace and stability. And it demands consultation and close cooperation with countries in the area which might be threatened.

Meeting this challenge will take national will, diplomatic and political wisdom, economic sacrifice, and, of course, military capability. We must call on the best that is in us to preserve the security of this crucial region.

Let our position be absolutely clear: An attempt by any outside force to gain control of the Persian Gulf region will be regarded as an assault on the vital interests of the United States of America, and such an assault will be repelled by any means necessary, including military force.

During the past 3 years, you have joined with me to improve our own security and the prospects for peace, not only in the vital oil-producing area of the Persian Gulf region but around the world. We've increased annually our real commitment for defense, and we will sustain this increase of effort throughout the Five Year Defense Program. It's imperative that Congress approve this strong defense budget for 1981, encompassing a 5-percent real growth in authorizations, without any reduction.

We are also improving our capability to deploy U.S. military forces rapidly to distant areas. We've helped to strengthen NATO and our other alliances, and recently we and other NATO members have decided to develop and to deploy modernized, intermediate-range nuclear forces to meet an unwarranted and increased threat from the nuclear weapons of the Soviet Union.

7. At the extreme, a portion of the cost of the Iraq war could be "allocated" in part to oil dependence. See, for example, Joseph Stiglitz and Linda J. Bilmes, *The Three Trillion Dollar War: The True Cost of the Iraq Conflict* (W. W. Norton, 2008).

8. Material in this and the following section is drawn from a paper, *Oil and Gas Security Issues,* I prepared for a project for Resources for the Future—*Towards a New National Energy Policy: Assessing the Options*—carried out in collaboration with the Tulsa-based National Energy Policy Institute (NEPI) supported by the George Kaiser Family Foundation, September 2009.

9. See table A11, EIA, *AEO 2010,* http://www.eia.doe.gov/oiaf/aeo/.

10. See EIA, "Assumptions to the Annual Energy Outlook (2010), NEMS." This report summarizes the major assumptions used in the EIA's National Energy Modeling System (NEMS) to generate the *AEO 2010* projections, http://www.eia.doe.gov/oiaf/aeo/assumption/index.html.

11. R. James Woolsey, "Turning Oil into Salt," *National Review*, September 25, 2007, at http://www.nationalreview.com/articles/222256/turning-oil-salt/r-james-woolsey.

12. An import tax is mentioned here as a possibility that is sometimes considered. I do not favor an import tax, because it restrains free trade and almost certainly would induce prompt and stern retaliation.

13. Of course, improved fuel efficiency will to some degree encourage more driving, because fuel will cost less, thus offsetting the decrease in consumption.

14. George Kaiser of the National Energy Policy Institute adopts a different line of reasoning to arrive at a level of imports that is sufficiently low to remove the ability of foreign oil producers to influence the U.S. economy and foreign policy. Kaiser proposes reducing global demand for imported oil in order to open a gap between global production capacity and demand, thus reducing cartel power. Kaiser assumes the U.S. target for reduction of demand for imported oil as 4 million barrels per day, achieved in stages, from the present level, and he argues that this amount of demand reduction is possible.

4. Biomass, Solar, and Nuclear Energy, with an Aside on Natural Gas

1. Pimentel's views can be found in D. Pimentel et al., "Environmental and Social Costs of Biomass Energy," *BioScience* 34, no. 89 (1984); and D. Pimentel, "Ethanol Fuels: Energy Balance, Economics," *National Resources Research* 12, no. 127 (2003).

2. *BTUs* stands for British Thermal Units, a measure of heat. One BTU is the heat required to raise the temperature of one pound of water one degree Fahrenheit.

3. In this analysis, natural gas is considered to be equivalent to oil (on an energy equivalent basis) because natural gas is substitute for diesel oil as a fuel for turbines in the industrial sector.

4. GAO Comptroller General Report, *Conduct of DOE's Study Group: Issues and Observations,* EMD 80–128, September 30, 1980. Available at http://archive.gao.gov/f0202/113655.pdf.

5. OTA, *Gasohol: A Technical Memorandum* (September 1979), available at http://www.princeton.edu/~ota/disk3/1979/7908_n.html.

6. H. Shapouri, J. A. Duffield, and M. Wong, *The Energy Balance of Corn Ethanol: An Update,* Agricultural Economic Report #814 (U.S. Department of Agriculture, July 2002). The data is taken from table 6, p. 9.

7. http://www.eia.doe.gov/oiaf/servicerpt/jeffords/.

8. Table A2 of the EIA *2010 Annual Energy Outlook* early release.

9. For a review of cellulosic biomass, see L. R. Lynd, "Overview and Evaluation of Fuel Ethanol from Cellulosic Biomass: Technology, Economics, the Environment, and Policy," *Annual Review of Energy and the Environment* 21, no. 403 (1996).

10. The DOE roadmap for cellulosic ethanol can be found at http://genomicscience.energy.gov/biofuels/b2bworkshop.shtml.

11. The DOE press release and a list of the projects are available at http://www.brdisolutions.com/default.aspx.

12. N. W. Lewis and D. G. Nocera, "Powering the Planet: Chemical Challenges of Solar Energy Utilization," *Proceedings of the National Academy of Sciences* 103, no. 15,729 (2006), available at http://www.pnas.org/content/103/43/15729.full.pdf+html.

13. There are many ways to estimate C. A useful formula was put forward in the American Physical Society 1980 photovoltaics study. See Henry Ehrenreich and J. H. Martin, "Solar Photovoltaic Energy," *Physics Today* (September 1979).

14. A simplified formula for the LCOE is:

$$\text{LCOE cents} / \text{kWe} - \text{h} = r(\text{interest } y^{-1}) \ 100C \ (\$/W_p)/(365d/y)$$
$$q(kWe - h)/(d - kW_p).$$

15. $1/watt PV DOE Workshop Summary, April, 2010, available at http://www1.eere.energy.gov/solar/sunshot/pdfs/dpw_summary.pdf.

16. DOE Photovoltaics Technology Roadmap 2003–2007, at http://www.nrel.gov/docs/fy04osti/33875.pdf. See also the International Energy Agency technology road map, available at http://www.iea.org/papers/2010/pv_roadmap.pdf.

17. Available at http://web.mit.edu/nuclearpower/.

18. Update of the 2003 MIT Future of Nuclear Power, 2009, available at http://web.mit.edu/nuclearpower/pdf/nuclearpower-update2009.pdf.

19. Lewis Strauss, chairman of U.S. Atomic Energy Commission, in a 1954 speech to the National Association of Science Writers: "It is not too much to expect that our children will enjoy electrical energy in their homes too cheap to meter." This prediction stands unchallenged as hubris of technologists who see the benefits but not the challenges of applying technology.

20. Two websites that offer reliable information and analysis of the Japanese nuclear accident are http://www.iaea.org/newscenter/news/tsunamiupdate01.html and http://web.mit.edu/nse/.

21. This is the overnight cost in 2007 dollars, i.e., excluding interest accrual during the construction period. See Y. Du and J. E. Parsons, *Update on the Cost of Nuclear Power,* MIT Center for Energy and Environmental Policy Research, Working Paper 09-004 (May 2009), available at http://web.mit.edu/ceepr/www/publications/workingpapers/2009-004.pdf.

22. Loan guarantees were authorized by Title XII of the Energy Policy Act of 2005 at the level of $8 billion. This amount was subsequently increased in the 2009 stimulus package.

23. The DOE press release is available at https://lpo.energy.gov/?p=817.

24. This point is discussed in the MIT Future of Nuclear Power study; DOE Photovoltaics Technology Roadmap 2003–2007, at http://www.nrel.gov/docs/fy04osti/33875.pdf. See also the International Energy Agency technology road map, available at http://www.iea.org/papers/2010/pv_roadmap.pdf.

25. http://www.eia.doe.gov/cneaf/nuclear/page/analysis/nuclear power.html.

26. Acknowledged nuclear weapon states are Britain, China, France, India, Pakistan, Russia, the United States, and North Korea. Israel is an unacknowledged nuclear weapon state. South Africa, Belarus, Kazakhstan, and Ukraine abandoned nuclear weapons upon the breakup of the Soviet Union.

27. "Global Energy Security," G-8 Summit 2006, St. Petersburg, Russia, July 16, 2006, item no. 31, http://en.g8russia.ru/docs/11.html.

28. John Deutch, Arnold Kanter, Ernest Moniz, and Daniel Poneman, "Making the World Safe for Nuclear Energy," *Survival* 46, no. 4 (Winter 2004–2005).

29. See the speech by former senator and Nuclear Threat Initiative co-chair Sam Nunn and NTI's press release, September 19, 2006, http://www.nti.org/c_press/release_IAEA_fuelbank_091906.pdf.

30. For a description of Iran's nuclear activity, see "Iran Profile," NTI, Washington, D.C., http://nti.org/e_research/profiles/Iran/index.html.

31. For background, see "Brazil Profile," NTI, Washington, D.C., http://nti.org/e_research/profiles/Brazil/index.html; see also Sharon Squassoni and David Fite, "Brazil as a Litmus Test: Resende and Restrictions on Uranium Enrichment," *Arms Control Today* (October 2005), www.armscontrol.org/act/2005_10/Oct-Brazil.asp.

32. For a description of the provisions of the act, see the fact sheet from the Office of the White House Press Secretary, December 16, 2006, www.whitehouse.gov/news/releases/2006/12/20061218-2.html.

33. India has three reprocessing plants in operation at Trombay (0.1 MT/year), Tarapur (150 MT/year), and Kalpakkam (100 MT/year). See http://www.globalsecurity.org/wmd/world/india/. Although India has agreed to place its civilian nuclear facilities under IAEA safeguards as part of the 2006 U.S.–India cooperative agreement, it is unclear which of these facilities will be classified as part of the civilian or military nuclear program. India is believed to have separated enough plutonium for many nuclear weapons.

5. Managing Energy Technology Innovation

1. Data from the National Science Foundation, *Federal Funds for R&D 2009,* available at http://www.nsf.gov/statistics/fedfunds/.

2. Although defense R&D is undertaken for military applications, there is a long history of exceptionally valuable transitions to the commercial sector—notably in electronics, information technology, the Internet, and materials.

3. The Energy Security Act of 1980 [S.932 Public Law 96-294 (06/30/80)] contains much more than just the creation of the SFC. It contained "something for everyone" (funded from the windfall profits tax), which explains why it passed. For example, it was likely the first legislation to authorize and fund a study of the climate effects of greenhouse gases: Title VII, Subtitle B, "Carbon Dioxide," directed the director of the Office of Science and Technology Policy to enter into an agreement with the National Academy of Sciences to carry out a comprehensive study of the projected impact on the level of carbon dioxide in the atmosphere of fossil fuel combustion, coal conversion, and related synthetic fuels activities. The law required a report with recommendations, to be submitted to Congress.

4. *Energy Projections to the Year 2000: July 1982 Update,* DOE/PE-0029/1. This document projected a range of 130–169 GWe U.S. nuclear power capacity in the year 2000; in fact U.S. capacity turned out to be about 100 GWe.

5. Termination of United States Synthetic Fuels Corporation Act, April 7, 1986, P.L. 99-272, Title VII, Subtitle E, 100 Stat. 143.

6. The large DOE synfuels demonstration plants—Exxon Donor Solvent and Solvent Refined Coal I and II—were terminated in 1981 and 1982 after vast expenditures.

7. A thorough discussion of the alternatives has been given by my colleague Richard Lester in his *America's Energy Innovation Problem (and How to Fix It): A Report from the Energy Innovation Project,* MIT Industrial Performance Center (2009), http://web.mit.edu/ipc/publications/pdf/09-007.pdf.

8. http://www.lgprogram.energy.gov/.

9. Paul M. Romer, "Implementing a National Technology Strategy with Self-Organizing Investment Boards," p. 345, in *Brookings Papers on Economic Activity, Microeconomics 1993:2*, ed. Marin N. Baily and Peter C. Reiss (Brookings Institution Press, 1993). (I thank my colleague Richard Lester for pointing out this interesting proposal to me.)

10. The Center for American Progress, an organization with which I am closely associated, is a leading proponent of energy banks. See John Podesta and Karen Kornbluh, *The Green Bank: Financing the Transition to a Low-Carbon Economy Requires Targeted Financing to Encourage Private-Sector Participation* (May 21, 2009), available at http://www.americanprogress.org/issues/2009/05/green_bank. html. A primer, *What Is the Green Bank?* (June 16, 2009), is available at http://www.americanprogressaction.org/issues/2009/06/green_ bank_primer.html.

11. http://arpa-e.energy.gov/Home.aspx.

12. I have long favored this approach to implementing government technology demonstration programs for commercial applications. In 1991 a panel titled *The Government Role in Civilian Technology* of the National Research Council (Board on Science, Technology, and Economic Policy) made a similar recommendation for establishment of a Civilian Technology Corporation with a broader mandate to demonstrate technology based on R&D advances. See also Harold Brown, John Deutch, and Paul MacAvoy, "Priming the High-Tech Pump," *Washington Post,* April 9, 1992, p. A27. I believe this approach is especially suited to energy technology. See "What Should the Government Do to Encourage Technical Change in the Energy Sector?" *Chemical Technology* (February 2007), p. 16. See also P. Ogden, J. Podesta, and J. M. Deutch, "A New Strategy to Spur Energy Innovation," *Issues in Science and Technology* (Winter 2008), http://www.issues.org/24.2/ ogden.html.

6. RECOMMENDATIONS

1. EIA, *International Energy Outlook* (2009), available at http:// www.eia.doe.gov/oiaf/ieo/pdf/0484(2009).pdf.

2. John Deutch, "The Good News about Gas," *Foreign Affairs* (January/February 2011).

3. The California Air Resources Board's Zero Emission Vehicle Program is described at http://www.arb.ca.gov/msprog/zevprog/zevprog.htm.

4. An overview of NEMS can be found at http://www.eia.doe.gov/oiaf/aeo/overview/.

5. The President's Council of Advisors on Science and Technology November 2010 report that recommends the QER may be found at http://www.whitehouse.gov/sites/default/files/microsites/ostp/pcast-energy-tech-report.pdf.

6. The best statement I know of the tension between analysis and the political process is an essay by James R. Schlesinger, *Systems Analysis and the Political Process,* Rand P-3464 (June 1967).

INDEX